Praise for *26 S*

"Where others see only 'human error,' Rossana takes us by the hand and shows us what there is to learn and love instead. Speaking *for* the dead speaks as loudly for Cesare as it does for everyone who could have been in his shoes. It speaks for the past as much as for the future—*our* future, and the future of Cesare's legacy."
— Dr. Sidney Dekker author of *The Field Guide to Understanding Human Error*

"*26 Seconds* is at once a searing portrait of grief and a damning investigation of an industry that puts profit over safety at every turn. With an engineer's exactitude, Rossana D'Antonio exposes how blaming "pilot-error" is often a cover for corporate greed and systemic failures. Like so many plane crashes, the flight that killed D'Antonio's brother was not an accident—and this powerful book is an affecting guide to understanding why."
— Jessie Singer author of *There Are No Accidents: The Deadly Rise of Injury and Disaster—Who Profits and Who Pays the Price*

"D'Antonio takes you on a journey of discovery. *26 Seconds* is a story that blends the humanity following an aviation disaster, an obsessive quest for answers, finding purpose beyond tragedy, and a scathing rebuke on the weaknesses in our aviation industry. Heartbreaking and so timely."
— Gloria Feldt author of *No Excuses* and Co-Founder & President of Take the Lead

"*26 Seconds* is impossible to put down. You will read it in one sitting. I happily lost a night's sleep as my heart raced, tears stinging my eyes. Turning page-after-page, I couldn't read fast enough to see if Rossana's heroic search to find the "smoking gun" that would ultimately clear her brother's name would be uncovered by her ceaseless efforts to research each detail of the disaster."
— Monica Holloway author of *Driving with Dead People*

"This story will move you to tears and move you to action—which is to say that it does what a great memoir must do. With the heart of a sister, Rossana takes us inside the life of a brother who loved to fly and died doing it. With the mind of an activist and an engineer, she takes us inside the aviation industry that is far too quick to lay blame on the pilot in the cockpit rather than examine its own shortcomings. A riveting read—and an important one for anyone who steps on a plane."

—Jennie Nash, CEO, Author Accelerator

"Rossana's passion for public safety is an impressive legacy that honors not only her brother but also herself. A profound, and well-written journey of healing from a very personal tragedy."

—Bonnie Comfort, author of *Staying Married is the Hardest Part: A Memoir of Passion, Secrets and Sacrifice*

"*26 Seconds* is a deeply moving memoir about loss, resilience, and the search for truth. When Rossana's beloved brother is tragically killed in a devastating plane crash, her world shatters in an instant—just 26 seconds of chaos altering the course of her life forever. This important story illustrates how unimaginable pain can lead to purpose—and the power to make a difference. Somewhere up there Cesare is smiling."

—Deb Miller, author of *Forget the Fairy Take and Find Your Happiness*

"*26 Seconds* pulls you into a deep personal family crisis on a personal level, where the reader has an unquenchable thirst to get to the truth themselves. It also highlights the author's defining moment, when she will not let governments and large, influential businesses hide the inconvenient truth by placing blame on 'pilot error.' Because of her valor, she will have saved countless other lives, while also reframing her brother's death as a hero who saved 130 lives. My hat's off to Rossana for telling this important story!."

—Maria Lehman, PE Director, US Infrastructure at GHD and 2023 ASCE President

26 Seconds

Grief *and* Blame in the
Aftermath *of* Losing
My Brother *in a* Plane Crash

ROSSANA D'ANTONIO

SHE WRITES PRESS

Published 2025

Printed in the United States of America

Print ISBN: 978-1-64742-904-1
E-ISBN: 978-1-64742-905-8
Library of Congress Control Number: 2025900942

For information, address:
She Writes Press
1569 Solano Ave #546
Berkeley, CA 94707

Interior design and typeset by Katherine Lloyd, The DESK

She Writes Press is a division of SparkPoint Studio, LLC.

For Mom and Dad

And of course,

For Cesare

Contents

Author's Note

This is a memoir—a true story. My story. As such, it reflects my recollections of my lived experiences over time. Many of the early memories following the plane crash are seared in my mind, as if having my very own personal movie reel allowing me to relive it at a moment's notice. This story was born on May 30, 2008, while vacationing in Spain. That night, my travelogue, its cover stamped with vibrantly colored wildflowers and butterflies, its pages filled with glorious memories, instantly became one of many journals that chronicled the tragedy and its aftermath. I have relied on those personal journals where I documented the people, events, timelines, sights, sounds, and raw emotions. I supplemented all of it by consulting with several people who appear in the book for their perspective or to fill in the memory gaps.

In some cases, names and characteristics were modified to protect the privacy of others. In other cases, I omitted people from scenes but only when that omission had no bearing on the substance of the story.

Overall, I believe I have done my story justice.

"Three things cannot be long hidden:
the sun, the moon, and the truth."

—Buddha

Foreword

"Cesare was a great pilot. He was one of our best. If he couldn't nail the landing that day, none of us could have."

Thus says Johnny, a colleague pilot. You'll hear him speak by the time you're well into Rossana's *26 Seconds*.

In the wake of a fatal accident lots of people are ready to speak about the dead—in this case, the dead pilot, Rossana's brother Cesare. This can quickly turn into speaking *badly* about the dead. For example, an accident investigation board may conclude that a crash resulted from a *poorly managed*, non-stabilized approach that resulted in a high crew workload during a critical phase of flight; the crew's subsequent *failure* to detect and respond to the decreasing airspeed and impending onset of the warnings during the final approach despite numerous indications; the crew's *failure* to abort the unstabilized approach and initiate a go-around; and their *failure* to respond to warnings in a timely manner, which resulted in a stall and subsequent ground impact.

Of course, those who have followed the development of human factors and safety science over the past decades will recognize this as "the old view," which is all about some person's failures and poor work, about "human error" being the cause of trouble in an otherwise safe system.

Being smart in hindsight offers those who speak *about* the dead plenty to go on. Find a series of mistakes, line them up, and trace the way to the bad outcome. It is so easy. The old view really has efficiency and expediency going for it. You don't have to do a lot of analytic work; you don't have to put in the hard investigative yards. Just say where you judge other people to have gone wrong (because you now know what the right thing was, given hindsight). And, by implication, suggest that had *you* been in that situation, you would have been quicker, smarter, wiser, safer . . . The vocabulary that remains is organized around human deficits. It is about their shortcomings, their delays and failures, their poor management and poor decisions. All this highlights, from the rubble of the accident, what they did not do.

Such an account about the dead is a ghost story and a lullaby all rolled into one. The account warns us of the evil sources of failure, and then it sings us to sleep with the lie that it's all under control. Such accounts and such investigations represent an ethical and intellectual shortcut. They ignore the host of complexly interacting factors that conspired against a good outcome, and that might do so again.

"Blame," said Jaime Goti, architect of trials against the military juntas that ruled Argentina in the 1970s, "simplifies reality by turning those blamed into the sufficient cause of a harm. This process implies removing from focus otherwise relevant contributions to the outcome."

Blame, in other words, is incapable of taking the non-deterministic, complex universe seriously. Blaming something or someone offers us an illusion of control over suffering and death.

But this is about human lives—particularly Cesare's—rich with experiences, expertise, professionalism, friendships, aspirations, intentions, hopes, loves, and dreams (and the assumption that he was going to turn around the airplane at some point and fly back, just like any other workday). Yet in hindsight, it is so very easy to formally reduce his life to a short paragraph of deficiencies. *That* is speaking badly about the dead.

Rossana does something very different in this book. In fact, she does the opposite. She doesn't speak *about* the dead—in a way she speaks *for* the dead. She beautifully and compassionately illuminates the hopes and constraints and obligations of Cesare's existence, his humanity, the patterns of cognition and performance he got configured into on the fateful flight. This is crucial. Because if what Cesare did—"one of our best," to speak with Johnny—made sense to him at the time, it will likely make sense to others as well. Rossana shows that our ethical, human (and humane, compassionate) commitment to speak *for* the dead irradiates beyond the dead. It applies to the future at least as much as it might help us unlock portions of the past.

Rossana's book is part of a particular genre in safety literature. In that genre, people speak of their lived experiences of a disaster, of getting caught up in the wash of a tragic event. As she offers us her story, authenticity is never in doubt: Rossana has lived out the full consequences of her own exposition. Remarkably, this genre has remained relatively stable over time. Plutarch's and Cicero's accounts of the lived experiences of losing their respective daughters some two millennia ago would seem familiar to readers today. C. S. Lewis's *A Grief Observed*—from the 1960s—would in turn not be strange to Cicero, or to Rossana.

Yet even as it does so, Rossana's book offers more. By revisiting Cesare's story, she is able to let us make authentic contact with the man, her brother, his flight, his life. For us still lucky enough to be here, there is the possibility of learning from the past from this writing. Where others see only "human error," and learn nothing of value (they have their lullaby and ghost story nicely buttoned up, after all), Rossana takes us by the hand and shows us what there is to learn and love instead. Speaking *for* the dead speaks as loudly for Cesare as it does for everyone who could have been in his shoes. It speaks for the past as much as for the future—*our* future, and the future of Cesare's legacy.

—Dr. Sidney Dekker
Professor of Humanities and Social Science
Griffith University

Prologue

26 Seconds

0:26   The rear wheels of TACA Flight 390 touch airport Runway 02 with a slight bump and a loud screech. Rainwater sprays up off the grooveless asphalt.

0:24   The brakes of the Airbus 320 are applied, starting the plane's deceleration.

0:19   The aircraft's nose slices through the thick fog, and the front wheels meet the black pavement. Flaps are lowered and the resistance is palpable, the powerful forward force pulling passengers' bodies away from their seats.

0:17   The landing is smooth. The passengers applaud.

0:14   The jumbo jet's brakes are fully engaged, but the steel giant continues to barrel down the runway.

0:09   The long deceleration is underway, but the slowdown is not enough.

0:05   The jet engines howl, the brakes shriek, the overhead compartments rattle. The plane barrels to the borderline of the rain-soaked runway, where the asphalt ends, and the drop-off begins.

0:03   The runway ends. There is no more pavement beneath the wheels. The jet propels over a ravine, hurtling across the highway below.

0:02   The bird's metallic nose smashes into the embankment. Inside the cabin, *BOOM!*

0:00   Silence. Blackness.

I vault upward in a half-awakened confusion, heart thundering. Eyes wide open, my breathing fast and heavy, I rub my eyes and try to erase the images that just terrorized my unconsciousness. The digital clock on the nightstand reads 3:32 a.m. in blood-red light, and I ease my head back down to the pillow. Out of breath, I try to regain a sense of control. My eyes focus on the moonbeam creeping in through the bedroom window. I sigh and tell myself it's only a dream.

But it's more than a dream. It's a nightmare that keeps coming back. And it keeps coming back because it's real. Twenty-six seconds. Twenty-six seconds after which the magnificent and boisterous beating of Cesare's heart was forever muffled into a silence that now echoes across my every dark, dark night. Twenty-six seconds that thrust my life into a tailspin and launched me on a quest for the truth.

I vow to find out what really happened in those twenty-six seconds if it takes every second I have left on earth.

PART I
Villain or Hero?

"The bluebird carries the sky
on his back."
—Henry David Thoreau

1

Tragedy

May 30, 2008

Exhausted and happy, we returned to our quaint hotel a short block from Barcelona's lively La Rambla, Spain's most famous promenade. Freddie and I were in Spain celebrating our third wedding anniversary, which we'd toasted two days earlier over a bottle of local red wine produced along the coastal hills of the Catalonia region. We'd selected a quiet tapas bar for our farewell dinner and were seated at a candlelit table for two in the corner by a window, where we watched the bustle outside while embraced by the calm of the seascape canvas paintings indoors. After two glorious weeks in a country of flamenco dancers, Spanish guitar, sangria, and colorful outdoor festivals, we were deliciously spent and maybe a bit heavy-hearted to leave. Romance, art, culture, and architecture filled every day. Picasso and Miró had been our tour guides, their art depicting human beauty in all its imperfect forms. But it was the spirited colors and fantastical designs of the brilliant architect Antoni Gaudi that held us captive for days. His daring and whimsical multistory structures of inanimate stone-work seemed to take on life in the forms of dragons, gladiators, and budding flower gardens. Angular, skeletal buildings with undisciplined undulating floors and facades that seemed to melt hearkened to fun houses and seemed oddly misplaced in the mid-dle of the bustling city. As engineers and admirers of the father of modernism, my husband Freddie and I roamed Barcelona's vibrant streets mesmerized by that enchanted land that shouted "*Gaudi!*" all across the city.

I plopped myself on one of the twin beds we'd pushed together to create a proper bed for a loving couple. There was no way Freddie and I were going to sleep apart, so we repositioned and spooned and wiggled to navigate the crevasse up the middle of our makeshift queen bed. By then we'd adjusted to the Europeans' propensity for twin beds but couldn't understand how they thought a hard bed was the same as a firm one. Perhaps something was lost in translation.

I was exhausted from a day of walking and sightseeing. We'd spent a large part of the day in a mad-dash, last-minute shopping spree to find the perfect replica of Gaudi's mosaic chimney, a trademark of the region's architecture. When we finally found the ideal one, we didn't even calculate what it would cost in dollars—we happily grabbed it, knowing it would look striking sitting in our glass display case at home, surrounded by other family treasures. In the months that followed, the trinket served as a colorful, vivacious reminder of those glorious days in Spain when we'd been so playful and carefree. We followed that purchase by buying a Hard Rock Café T-shirt we'd agreed would be perfect for my brother, Cesare, and I couldn't wait to give it to him the next time he flew through LA. As with all things, he had no patience, so I knew he'd immediately pull it over his head and find the closest mirror to admire himself in his Spanish keepsake. The image of it made me smile.

I flopped back on the bed, reached for the TV's remote control, and turned on CNN. It was the only channel we'd been able to find in English, so for two weeks, it had been our link to home. On the screen were images of an airport, and a newscaster reported, "A commercial airplane has crashed in Honduras after overshooting the runway. The pilot and two passengers are confirmed dead." At first glance, the jumbo jet appeared to be intact, but when I looked closer, I could see something was terribly wrong. The fuselage had broken into three large segments, their edges crumpled like discarded sheets of paper. The jet's main body lay askew and rested unnaturally on its belly, its wings slightly sagging, as if defeated. The cockpit was crushed where it had come to a violent stop on a dirt embankment. As the camera panned across the crash site, I saw the initials *TACA* on the plane's blue tailfin.

TACA Airlines.

I felt hot lightning shoot through my stomach.

TACA Airlines was the flag carrier for El Salvador and one of the biggest airlines in Central and South America.

My brother, Cesare, was a pilot for TACA Airlines.

The scene on TV was chaotic. Throngs of people swarmed the plane, pulling injured passengers out. Ambulances with flashing red lights and wailing sirens were being loaded with people on stretchers. There appeared to be a crushed vehicle under the jet. I couldn't turn away from the TV. My brain was already tinkering with hypothetical root causes for the crash even as I battled the wave of anxiety and the air of fascination for the disaster before me.

Plane crashes had always fascinated me in a morbid way. Spain's Tenerife Airport disaster, the bombing of Pan Am's flight 103 over Lockerbie, Scotland, and the explosion of TWA flight 800 off the coast of New York had made significant impressions on me. I always fixated on the randomness of a plane crash, the extraordinarily bad luck of those who had boarded the wrong plane on a given day, and how in a flash of a moment, each of those catastrophic incidents had claimed hundreds of lives. It rattled me that all those human beings had boarded those particular flights when they might have selected any one of hundreds of other options that weren't destined for disaster. A roll of the dice. Suddenly, their time was up.

When my little brother Cesare had become a pilot in 1993, anxiety joined my disturbing attraction to these kinds of phenomena. After Cesare got his wings, at the news of any plane accident, I'd watch the story unfold, worried that Cesare might be involved even if I knew he was thousands of miles away from it. I'd sit glued to the TV waiting for details to emerge that would reveal my brother was safe. History had always proven my worry had been yet another instance of a protective sister clutching her pearls.

But this one was way too close to home, literally. Honduras, border country to El Salvador, where most of my family lived and out of which my pilot brother was based. And right there onscreen was the wreckage of a TACA Airlines plane. I grabbed my Black-berry and dialed Mom. The line was busy. *Damn!*

I wondered if Cesare knew the pilot, if he was a friend, if the pilot's family had heard the news. I wondered if my brother was somewhere suffering over the tragic loss of a colleague.

I could hear the faucet running in the bathroom as Freddie brushed his teeth. "Sweetie, it's a TACA flight," I murmured, try-ing to hide the uneasiness building inside me.

I dialed my mother's number again. Still busy. *Shit!*

Suddenly, Freddie's Blackberry buzzed to life, and he lunged toward the bed to grab it. My attention was on the TV, but I heard him say, "Hi, Andres."

My head snapped in Freddie's direction as he slowly sat on the bed, slightly crouched over. Why was Andres, my nephew, my sister's eldest son, calling Freddie? My insides were suddenly ablaze. A wave of burning heat raced through my chest and toward my extremities. "Oh my god, no. Oh my god, no. Oh my god, no!" was all I could utter.

Then my mind shot off into magical thoughts, as if this pan-icked chant could change the events I instinctively knew were unraveling before me. I could hear my heart pounding in my ear-drums, and I flailed my legs as if trying to fend off the inevitable reality that was trying to creep into my safe haven.

Freddie's face was a mask of artificial calm. "Yes, we're watching it on CNN. Has it been confirmed?"

"Oh my god, no. Oh my god, no. Oh my god, no!" I wailed. The images continued to flash on the screen. People swarming the broken plane. Chaos. A passenger freed from the wreckage, his face stamped with pain. Panic. Tears washing down people's faces. Heartbreak.

"Can I speak to Mami?" Freddie asked Andres. My husband had adopted my mom as his own and called her the Spanish version of *mommy*. I loved that. It was typical of his warmth and his love of family. *Mami* was probably the only Spanish word he knew, aside from the four-letter words Cesare had taught him. Quietly, Freddie said, "Okay, we'll call you later."

"Oh my god, no. Oh my god, no. Oh my god, no!"

Freddie dropped the phone, sat next to me on the bed, and wrapped his arms around me. But I couldn't feel him there. I couldn't feel anything, and all I could hear was the pounding of my heart. The pounding of my heart, which somehow told me I was alive, and Cesare was dead.

Freddie and I sat like that, oblivious of space and time. I stared at the TV, but the recycling images were now a blur.

Somehow, I was supposed to make it through the night. And the next day, my husband and I would have to get on a plane.

Then another. And another.

2

Finding Our Calling

As far back as I can remember, Cesare had wanted to be a pilot. Actually, at first I think he wanted to be a policeman. He was five years old the time he and I approached a policeman parked in a two-tone car on our street on the west side of Los Angeles.

"Can we get your autograph?" Cesare asked the officer.

The nice officer stepped out of the black-and-white, insisting he wasn't a famous actor like those on *Adam-12*, a popular cop show back then in the '70s, but we didn't care. I was nine, and I wanted his autograph because he was cute, but Cesare was in true awe as he stared up at the tall, lean officer with the shiny badge on his chest and the menacing gun in the holster on his hip. We finally wore the cute cop down, and he gave us his autograph, a bunch of scribbly marks on a little piece of paper.

"Ya'll go on home now," he said.

We did as we were told, turning to wave multiple times as the cute cop waved back, a big toothy smile on his face.

I lost interest in the police officer, but Cesare didn't; for years, he kept the paper with the scribbly marks in the top drawer of his nightstand. It was the same night table that held some of his other prized possessions, including his two favorite Matchbox cars, a handful of multicolored marbles, a few plastic green toy soldiers, and a pair of real drumsticks he got from who knows where.

I think all little boys want to be cops early on. It's the first visible influence of the media on kids: glamorous images of real-life heroes fighting the bad in the world in the name of justice for all. In fact, the power of the media is such that it can create heroes and villains with the flick of a pen. Where one falls on the hero/villain spectrum is based on who controls the bullhorn and the nature of the message the bullhorn holder delivers.

Cesarino, as we affectionately called him, had a quick, curious mind and like so many other boys, loved to play cops and robbers, cowboys and Indians, warriors and rebels. Of course, he always wanted to play the hero and would chase me, the self-declared villain, around the house. Sporting his Lone Ranger domino mask, Cesare would cock the barrel of his toy gun and shoot in my direction as I clambered over the living room furniture.

"Why are you still running?" he'd yell. "I just killed you!"

Cesare believed the good guy always won, and this led to much frustration on his part when I didn't abide by his unrealistic idea of how the world worked.

Although Cesare loved to play war, I don't think he ever really wanted to be a soldier. But he did have an official green military helmet my sister Marcella bought him from our local Army surplus store. When I was nine years old and Cesare was five, Marcella, who we all called Mars, was a sixteen-year-old, self-proclaimed hippie, the family artist who spent countless hours strumming her guitar, writing poetry, and composing songs. She also spent a lot of time at the surplus store shopping for her wardrobe essentials: Levi's blue jeans, oversized denim shirts, and colorful bandanas. The helmet was designed for a grown man or a fearless almost-man who thought he was immortal and was primed for the front lines. But on Cesare's little head, it wobbled and thumped as he ran rat-tat-tatting the enemy.

About that time, Cesare got an aviator's hat from the same surplus store my sister was helping to keep in business. It was leather and had earflaps, the kind of helmet worn during World

War I. It was meant to be worn with aviator goggles, so Cesare made do with swimming goggles. He ran around for hours with that goofy cap on his head and fluorescent red swimming goggles covering his eyes, his arms spread out wide, gliding like a make-believe airplane through the neighborhood streets on the west side of Los Angeles. "Look at me! I'm the Red Baron!" he squealed. At five years old, he'd already found his calling, and he never wavered from it.

When Cesare was ten years old, he became hooked on an Argentinian soap opera about a handsome airline pilot and his romantic exploits. Even the title, *Amandote* (Loving You), was alluring. The show depicted the glamorous life of a modern-day playboy who jet-setted around the world having romantic affairs with beautiful women in every port. Cesare was in awe of the main character and would often recite lines and reenact scenes from the latest episode.

Mom wasn't thrilled about the idea of her son becoming a pilot. She argued, "It's so dangerous. If God had intended for man to fly, he would have given him wings."

Even at age ten, Cesare was prepared with a rebuttal. "Mom, do you know that more people die in car crashes than in airplane accidents? Flying is the safest mode of travel."

For years as a successful pilot, he proved the old adage right. But after Cesare died, I looked up the statistics for his claim, wondering whether he'd had the facts straight or was simply parroting others.

According to the U.S. Department of Transportation, each year 1 out of 7700 drivers die in automobile accidents while 1 in 2.1 million people die in airplane accidents.

Holy crap, I thought, *one in 2.1 million!* Cesare had been right. After discovering that information, I wondered many times if my brother had ever imagined he'd become that one in 2.1 million.

One in 2.1 million. The statistical unlikelihood of my brother's death by plane turned out to be our unimaginable reality.

Cesare was determined to enroll in flight school, and as soon as he earned his private pilot license, he headed out for regular afternoon flights by himself to enjoy the solitude in the wide expanse of blue all around him. His free spirit thrived on the freedom only flying could give him, freedom without roads or structure or clearly defined lines.

If Cesare was the dreamer, I was the bookworm. Where he sought to explore the heavens, my feet were planted firmly on solid ground. By the time I was about ten years old, I was mesmerized by the beauty of regal mountain ranges and gently rolling hills. While riding in the back seat of our family's big Oldsmobile Cutlass, I gazed at their majestic peaks and low-lying valleys in the distance, tracing their smooth outlines with my index finger and somehow feeling a connection. There was something about towering mountains rising to touch the sky that gave me a sense of peace.

I loved to build castles out of mud and sand, carefully refining the water content and soil mixture like a mad scientist until I'd created a structure stable enough to stand on its own. I repurposed my Lincoln Logs and designed an intricate system of roadways and bridges specifically for our Matchbox cars. And I was especially fascinated by the most famous of civil engineering failures, the Leaning Tower of Pisa, so I spent endless hours reading and conjecturing what had caused the tilt. My childhood hypothesis was "quicksand, just like in the cowboy movies." I didn't know it then, but I was already showing clear signs I was meant to be an engineer.

But that wasn't my only calling.

I was four years old the day my parents brought Cesare home from the hospital. My mother handed me the baby bundle, and my father said, "Take care of your little brother. Always take care of your little brother."

This sense of love and protection for our little brother was borne from Dad's Italian culture. The strong bond of family, *la famiglia*, was paramount with him. As a young man, Dad had left his home in Rome following World War II in search of a better life in America. He arrived in the US with very little beyond strong family values that he had instilled in us. As Dad's namesake and the youngest, *il piccolo*, Cesare became the center of our nuclear family.

I looked down at the swaddled Cesare sleeping in my arms and felt tall with pride. I took that command very seriously and for the next few years rarely let him out of sight. I always alerted the adults when he wandered into the deep end of the pool, and I shooed away neighborhood bullies.

Then when I was thirteen and Cesare was nine, the unimaginable happened. Our father was diagnosed with terminal lung cancer. Our parents had chosen to shelter us from the truth, but it stumbled clumsily out into the open, as truth often does.

Dad asked me to go with him on an errand. Buzzing with excitement, I accepted the offer. It would just be the two of us! I jumped into the passenger seat of the Cutlass without asking where we were going, although anyplace would have been fine with me. I would have gone to the end of the world with my dad.

We arrived at a fancy office building in West Los Angeles, a relatively short drive from home. A smiling lady dressed in a pin-striped business suit greeted us at the door and invited us to sit, but I wandered over to the framed diplomas on the wall. As I tried to decipher the meanings of some of the words and phrases on the diplomas, the chitchat at my back became an unintelligible drone. Then I heard my father say something devastating, and he said it just as casually as someone ordering a pizza. "I have lung cancer," Dad said to the lady across the desk. The words sliced through me like ice water.

I stood frozen in place, afraid I'd just overheard something forbidden. *Cancer.* Did Dad know I'd overheard his conversation?

He'd only spoken the words after I walked away. Or was this his way of telling me?

I couldn't bring myself to look at my father. I didn't really know what cancer was, but somehow, I already knew Dad was going to die from it. Not knowing what to do, I chose to keep staring at the wall. Behind glass, the diplomas now seemed to streak black ink down the cream-colored background.

The drive home was quiet. I expected Dad to console me, to tell me all would be well. He didn't. Maybe he expected me to ask about what I'd overheard. I didn't.

I couldn't tell anything was wrong with my dad, so I reasoned I'd been mistaken, that Dad wasn't going to die after all, and we'd continue to live together as a family until happily ever after. But several weeks later, cancer, the beast, came calling.

One Saturday night, we rushed Dad to the hospital in a mad panic as he struggled to breathe. He shuffled into the ER, steadied by Mars, every step an agonizing gasp for air as he seemed to plead for help with wide-open eyes. Several nurses wheeled him away, slapping an oxygen mask over his face as they hustled down the corridor, Mom and Mars running alongside. Cesare and I ran behind, two of our steps for every one of the adults'.

As the group turned the corner and disappeared, my brother and I stood, relatively lost in the hospital. Fear was stamped on Cesare's face, and in my head, I heard, "Take care of your little brother." The words compelled me to muster all my courage.

For the next couple of hours, Cesare and I roamed the hospital lobby trying to keep ourselves entertained. The TV played the news, which we had no interest in, and all the hospital magazines, raggedy with age, were filled with makeup ads and cake recipes. We walked through the wide hallways until we stumbled upon a clay model of the hospital with little cars, tiny people, and miniature trees. In that make-believe world, everything seemed frozen in time. I wondered if Cesare noticed that in the mini medical world on display, nobody was rushing around in a panic, the tiny people weren't wearing oxygen masks, the sirens of the little ambulance didn't appear to be wailing, and the teeny traffic light never turned green. As we stood there staring at the miniature campus, I focused on the bright red letters that spelled *EMER-GENCY* in the corner of the model. And I prayed the doctors in that ward could fix my daddy.

Suddenly, Cesare covered his face with the crook of his arm and started to cry. "Daddy's going to die, isn't he?"

The question shocked me. Cesare and I had never talked about Dad's illness. I was just a kid myself, and now Cesare had cornered me with the most serious question of his short little life.

Take care of your little brother. Always take care of your little brother.

I knew he deserved the truth. He was counting on me for answers. I turned to face him and looked him straight in the eyes. "Of course not. Don't be stupid. He's just not feeling too good right now. Don't cry, he's gonna be just fine." At thirteen years old, I hadn't yet learned how to show much sensitivity, so I threw a quick arm around Cesare and jostled him a little. Then I wondered whether he actually believed me. And I wondered if the burning in my stomach was from knowing the truth, that Dad would die soon, or the fact that in my desire to protect my kid brother, I had actually betrayed him.

Dad remained in the hospital for a week before coming home. There was really nothing else that could be done, the doctors said. Over the next several months, Dad's health worsened. When Cesare and I got home from school, he greeted us with a sad smile, his eyes tired and hollow. We now had oxygen tanks delivered to the house to help him do the job his lungs could no longer do.

One rainy afternoon, Cesare and I were at the dining room table doing our homework, although distracted by Dad's heavy breathing in the other room. *Wheeze.* By that time, his breathing was a series of deep gasps followed by a short pause and then another deep inhale. I'd started to time the pauses between the gasps, during which I stared blankly at the words in the textbook while concentrating only on the jagged rhythm of my father's breathing. Then the pauses between his breaths grew strangely long. I looked up from the book and saw Cesare staring at me, fear stamped on his face. I shot up from the chair and ran the short distance to Dad's room only to hear him take a deep gulp of air. I shuffled back to the table and smiled a tight-lipped smile at my brother. There were no words spoken between us, but we knew the end was near. Although cancer consumed his body, Dad's spirit remained steadfast. If he was scared, I never saw evidence of it. "The only way to overcome fear is to face it," he'd say in his strong Italian accent, and we were witness to his fearless battle with the beast.

I no longer prayed for Dad to get better. I prayed for no more suffering. That night, my prayer was answered. Just after midnight, Dad took his last breath.

Our father died when Cesare was only nine, and I've always wondered if the jolt of Dad's early death propelled my brother to live his life with urgency. For Cesare, flying was much more than a career choice; it was the path to a way of life he needed. Personally, geographically, and professionally, he couldn't be limited to a confined space with rigid boundaries. As a kid, when Cesare had been given a box of crayons, he reached for the flashy colors. When given the chance to paint, he pushed his brushstrokes to the far corners of the paper. So it didn't surprise me Cesare lived his life guided by a map of his own design. He created that map using those vibrant colors and wide brushstrokes, exploring the four corners of the world. For fifteen years, he lived out his dream.

When I heard my brother had died, I remembered my father's words. *Take care of your little brother. Always take care of your little brother.*

Back then it had been a confusing assignment, but after the crash, it became my mission. Days after the crash, in a fog of helplessness and grief, I turned to the only source of comfort that promised long-term relief: facts. I decided to unearth every last detail I could about what happened the day my extremely competent, highly focused, and expertly trained pilot brother had died in a plane crash.

I'd read about countless plane crashes, so after the crash of TACA 390, I knew it wouldn't be long before I'd start reading the words *pilot error*. Many people would be motivated to tarnish Cesare's professional reputation and draw the focus away from other possible reasons for the crash. I had to protect Cesare. I had to take care of my little brother.

3

Getting Home

The reality of the crash was so big, so incomprehensible, that my mind and body nearly shut down. That first night, still in Barcelona, I lay in bed, blanketed in numbness as I quietly watched the recycling of the news in a futile attempt to understand. Maybe if I heard the worst part enough times, I could believe it: Cesare was gone.

Cesare was gone?

Cesare was gone. How was that possible?

After Freddie and I said we should try to sleep, after we'd turned off the lights and lay awake on the first night of our new reality, I heard Freddie crying next to me. Like a robot, I laid my hand on his shoulder to comfort him. Unlike Freddie, I had yet to shed a tear, but my husband let his emotions flow freely.

Freddie had always loved my brother as his own, and the two of them had grown close in the few years they'd known each other. Whenever Cesare flew into Los Angeles, Freddie always wanted to be part of our jam-packed days of shopping and dining, and he loved to tell the story of the time Cesare shopped for shoes on Rodeo Drive in Beverly Hills. After trying on several pairs at the Gucci store, walking up and down the length of the shop to ensure the fit and look were just right, Cesare finally settled on two pairs. When Freddie heard how expensive the shoes were, he gasped,

"Six hundred dollars?! For *each* pair? Are you nuts? That's more than I paid for all the shoes in my closet!"

Looking down at Freddie's scuffed-up sneakers, Cesare laughed his signature, part-cough giggle. "I wouldn't go around bragging about that."

In the darkness of the hotel room, Freddie squeezed my hand, and his hushed whimpering reminded me that he, too, had lost his little brother.

On that first night, I tortured myself with thoughts of Cesare's final moments. I wondered what he'd been thinking. If he'd said anything. If he'd shouted. If he'd felt panic. If he'd died quickly. If he'd felt pain.

When had I last seen him? When had I last heard his voice? I couldn't remember. Sometime in the middle of the night, I gave up trying to sleep, got out of bed, and e-mailed my boss in Los Angeles with the news. I was usually lightning fast with emails, but now my fingers dragged over the keys trying to make sense using the scattered letters on the keyboard. Finally, my trembling fingers managed to finalize the message: *There's been a plane accident and my brother didn't make it. We are flying to Los Angeles tomorrow as scheduled but have arranged a connecting flight to El Salvador to be with family. I will keep you posted.*

As a civil engineer at Los Angeles County Public Works for almost twenty years, I had colleagues who were like family, and they all knew Cesare was a pilot. As far as I knew, they'd already read about the crash in the *Los Angeles Times*, had already seen the horrific picture of the broken plane on the front page. Everyone in my life would know my only brother had been killed, words I couldn't bring myself to type yet, let alone say out loud.

Freddie and I survived that first night and somehow managed to pack our bags, check out of our hotel, hail a taxi, ride to the airport, and board a plane back to the US. We made it through all those steps like a couple of robots. Our first stop would be Newark, for a three-hour layover. Then we'd catch a connecting flight to Los Angeles and then another flight to be with my family in El Salvador. There was just no simple, direct way to get from Barcelona to San Salvador. It was hard to believe this was our best option. But there were moments when I was almost grateful for the grief. In the days that followed the crash, I was so weary and confused and miserable, it barely occurred to me that immediately following my brother's horrific death by plane crash I'd be up in the air on at least three flights in twenty-four hours.

As Freddie and I waited to board our plane to Newark, I struggled just to hold up my heavy head with my tense, weary neck. I hadn't slept in almost thirty-six hours. We sat on the plane for nine hours from Barcelona to Newark, and I don't remember even a minute of the flight. Nine hours erased from my life. I have no memory of boarding the plane, of where I sat, of any interaction with a flight attendant. Did I carry luggage? Did I speak to anyone? Did I talk to Freddie? Did he hold my hand? Did I clutch the armrest in fear all nine hours? I don't know. Those nine hours remain a void, a black hole, as if marking the end of an era and the start of a new age, the beginning of Life Without Cesare.

I do remember arriving at Newark airport and being herded through customs along with countless other passengers who looked tense and exhausted. What were their stories? And what would they think if they knew mine, my real-life tale of my pilot brother who'd died in a plane crash just one day earlier? Would they consider that an omen? If any of them were booked on our flight to LA, would they dash to a ticket agent and change their plans?

After clearing customs, Freddie and I plodded through the airport, and I caught sight of a flight information screen. I looked down the list to confirm the gate number for our next flight and came to an abrupt stop on *Delayed* in bold red letters. There were no specifics, no explanations, no projected takeoff time—only the word *Delayed*. It almost seemed as if the travel gods were determined to prolong our already grueling voyage. This lack of control sparked a wave of anger in me. *Why the delay?* I wondered. *Poor weather? Aircraft malfunction? Plane crash?*

My thoughts raced back through the collection of plane crashes in my mind. Wasn't Newark Airport the site of a tragic plane crash years earlier? Maybe in the 1990s? Wasn't it a Federal Express cargo plane that had crashed while trying to land? Didn't it flip over and catch fire? That happened, didn't it? Or was I making that one up? Imagining things? Losing my mind? Wait, maybe I was thinking of the Tom Hanks movie, *Cast Away*. Wasn't that a FedEx crash?

I shook my head, closed my eyes, and tried to calm my mind. The last thing my psyche needed was to reason through plane crashes. I told myself to just focus on getting home, then to my family.

It was now midnight. I hoped our delay wouldn't be a long one, that we'd have plenty of time to get from LAX to the Van Nuys long-term parking lot, then home to Porter Ranch in the San Fernando Valley and back to the airport in plenty of time for our 1 p.m. flight to El Salvador.

We settled into a couple of seats at the gate of our delayed flight, and I thought about the fact that my brother had just died in a horrible crash of a huge airplane and here I was, only thirty-six hours later, about to board my second jumbo jet today. I'd just flown nine hours and was about to get on yet another airplane to fly thousands more miles over land and sea. My mind reverted to magical thinking. Today, I was probably safer than usual, right? What were the odds a second plane carrying a member of my family would crash within two days of the first? Didn't people often reason that a plane crash today made them safer fliers tomorrow? The odds were in our favor, weren't they? I tried to recall anything from my college statistics class that could support this reasoning. But that day, statistical analysis was too much for me. I'd already overtaxed my brain, and the fog settled back in.

I got up and walked around the airport food court looking at my options. I'd always avoided airport food, even under the best of circumstances. There was greasy pizza, dry sandwiches, MSG-riddled Chinese, and stale baked goods. It all made my stomach turn. I settled on an overpriced bottle of lukewarm water and popped two Tylenols. I'd lost count of how many I'd taken since hearing the news.

It was now Saturday night of Memorial Day weekend, and Newark Liberty International Airport was busy with travelers. I sat sipping my water and thought about the meaning of Memorial Day, to honor those who died serving our country. For me, it had always just been a three-day weekend full of barbecues and pool-side fun. The unofficial start of summer, the smell of suntan lotion and charbroiled hamburgers. But suddenly in the airport, I felt a deep connection to the holiday. In a way, Cesare had died in the line of duty.

We sat at the gate, Freddie holding my hand, and I thought about my conversation with my sister, Mars, the night before. She was the sensitive child. She cried easily and allowed her heart to lead her decision-making. I, on the other hand, have always had a keen ability to step outside myself and analyze a situation objectively. I can distance myself from a personal crisis and focus on what needs to be done. I can leave emotion out of it, excuse it from the room, send it to a cozy, quiet place out of the way so I can stay in command of my detailed to-do checklists and deadline-riddled spreadsheets. I'm sure I'm often mistaken for cold and unfeeling, but I consider myself practical and pragmatic. When there's a crisis, I'm the one you want on your team.

Mars picked up the phone on the first ring. "Hey . . ."

I blurted, "How's Mom?"

Our mother had suffered a heart attack eight years earlier, and every day since then, I'd worried that something would set her off, so much so that I called to check in on her every day.

"She's devastated," Mars said between sobs. "But the doctors are here, and they're looking after her."

Okay, Mom had doctors looking after her. Check. On to the next matter of business. "How's Cesare getting home? Should I make arrangements to fly down and pick him up?" *Take care of your little brother. Always take care of your little brother.*

As Mars spoke, her hushed voice cracked. "The airline is making all the arrangements. He'll be home tomorrow at the latest." There was a long pause, then she added, "Just come home as soon as possible."

"I will," I muttered.

"I love you," she said, openly sobbing again.

"I love you too."

When was the last time I'd told Cesare I loved him? I couldn't remember.

Since hearing my only brother was gone, I still hadn't cried, but as I sat at the gate at Newark's busy airport, I could feel tears well up inside, starting to heat up my face. The tears were coming, and I could tell there would be no holding them back. I hadn't taken off my sunglasses since leaving the hotel in Barcelona, and I was grateful for their cover as I rushed to the nearest restroom. The empty bathroom was an assault of fluorescent lighting and ammonia. I dashed into a stall and closed the door behind me just in time for the wave to hit. Leaning against the cold door, I was overcome by racking sobs and uncontrollable trembling throughout my body. As I stood there trying to muffle the sobs with the palm of my hand, I kept telling myself to stop. *STOP IT! Just stop it! Keep it together. You can't fall apart—there's too much to do. Focus!*

But the tears didn't care about my need for control. I stood in the stall and sobbed. It was an explosive moment of grief as streams poured from my eyes and my body quaked until all the tears I'd kept dammed gushed out of me. I cried so hard I couldn't see a thing, my vision a flooded blur. I steadied myself against the stall door spent and exhausted, then shuffled out of the stall, leaned on the wash basin, and splashed cool water on my flushed face. I stared at my image in the mirror and hardly recognized myself. My reflection was drawn and haggard, the circles under my eyes murky and dark. Where had the real me gone?

I dragged myself back to the gate and sat next to Freddie in a far-off corner where we'd been lucky enough to find two relatively isolated seats. Once again trying to hide behind my sunglasses, I hoped no one had noticed the wrecked woman who had shuffled out of the restroom only to flop like a rag doll next to a sad, tired-looking man. But it seemed no one saw me. They were all caught up in their own lives and were probably unaware of their blind trust that the planes we were about to board would deliver us all halfway across the world without incident.

A tall, lanky, blond-haired boy shuffled through his backpack looking for something he couldn't seem to find. A few chairs over, a young girl wearing headphones sat with her eyes closed, apparently lulled to sleep by whatever played on her phone. An old lady sitting across from us was engrossed in a paperback. All around me, people functioned normally, people whose lives hadn't been shattered. I was in a room full of people, but I was alone.

After being delayed for three hours, Freddie and I boarded our flight to Los Angeles. We would now arrive in LA at 3 a.m. rather than at midnight as planned. That would leave us about seven hours to get home, do laundry, repack our bags, and head back to the airport for a 1 p.m. flight to El Salvador.

I sat in the dreaded middle seat and as the lights in the cabin were turned off, I closed my bloodshot eyes to soothe the burning. As I drifted into semi-consciousness, I was sure I felt us hit an air pocket, and suddenly I heard a loud explosion in the front of the plane. A hot, suffocating ball of fire raced down the aisle, orange and black waves of gases scorching everything in its path. The overhead compartments rattled open and carry-on bags flew into the fireball. The inferno reached us and—

I jolted to attention with eyes wide open and let out a gasp. I whipped my head from side to side, but there was no fireball. There was no pressure pinning me to my seat.

Freddie slept peacefully next to me. We were still in the air, 35,000 feet over Mother Earth. We were safe.

Safe. Because I trusted the pilot who navigated this big, bold flying machine. A pilot who, like Cesare, loved the profession, perfected critical skills through training, and implemented safe flying practices. But I knew it wasn't just about the pilot. Pilots are spokes in a wheel. A pilot is only one player on a team. Every member of the complex machine that makes up a single flight has specific roles and responsibilities to keep the system working. I trusted that in a few hours we'd land on a runway that was well-designed, well-constructed, and well-maintained. I trusted there were safety policies and comprehensive training programs in place that all airline and airport employees adhered to strictly. I trusted regulators ensured compliance with all safety requirements. There were countless points of trust involved in air travel, issues most air travelers probably never think about.

But I did.

Trying to keep my mind from wandering into doomsday scenarios, I started a mental list of things I'd need to do in the seven hours we'd have to work with between flights. During that time, I planned to unpack and wash our clothes, sort through the mail, water the plants, repack for the funeral, and maybe, just maybe, get some rest.

We arrived at LAX at 3 a.m., physically and emotionally weary, still in shock, and three hours behind schedule. I sat on the cold long-term parking shuttle bench waiting for the rambling bus that would get us one step closer to my brother. We'd left our car in the Van Nuys FlyAway long-term parking lot, about twenty miles from LAX and a fifteen-minute drive from our home. Freddie paced back and forth on the airport median, the strip of concrete intended to maintain traffic flow by keeping the shuttles away from the arrival area. We waited half an hour for a shuttle that typically ran every ten minutes. I sat shivering and staring into nothingness, just waiting, and was jolted back to attention at the sound of Freddie's voice. He was talking to the driver of a shuttle heading toward Union Station. "Do you know where the Van Nuys shuttle is? We've been waiting for over half an hour now." His question was more a plea for help than a query.

The driver answered, "You didn't hear? There's been a shooting in Van Nuys. The whole area is cordoned off. No buses in or out."

I shook my head slowly and squeezed my eyes shut in frustration.

"But our car is at the Van Nuys parking lot," Freddie insisted, as if this explanation might somehow change anything.

"Your best bet is to get a taxi home, then pick up your car in the morning. You're not getting in there tonight, I can tell you that." I wondered if the driver would have given us a different answer if he'd known what we were going through.

"Thanks," Freddie said softly and turned to look at me, his shoulders slumped and eyes downcast as the bus rambled away.

Rage rushed into my veins, my lungs, my cells, and I was tired of trying to contain it. I thought, *Fuck you, God!! Are you fucking kidding me?! What more are you going to send our way, you asshole?!* I wanted to lash out, to hit something. Instead, I covered my face in my hands and sat there exhausted and defeated.

We hailed a taxi and rode through the city's half-deserted freeways and streets in silence. I sat and shivered until Freddie reached for my hand and gave it a soft squeeze. I rested my head against the window and watched the blur of the city lights. The ticking of the meter indicated we were steadily making our way home, a few dollars at a time. Finally, the taxi turned a corner and stopped in front of our house, which now seemed slightly unfamiliar. I looked at my watch. It was after 4:30 a.m., approximately forty-four hours after Cesare died.

Forty-five hours ago, my brother was alive.

He'd been alive on this earth only two days earlier.

I wondered how long I'd be able to use math to try to make the loss seem just a bit less horrible. A loss I could still measure in days.

4

Limbo

My knuckles were white as I clutched the receiver. "*What* are you talking about?"

It was 9:30 a.m., and we'd been home for five hours when a TACA Airlines representative called to tell us we'd missed our flight to El Salvador. I was disoriented, exhausted, and irritated, but I tried to maintain my composure. "We're getting ready to go to the airport now. We're on a 1 p.m. flight. We have plenty of time," I said, glancing at my watch.

The voice on the other end was soft and kind. He knew exactly who we were. TACA staff had also suffered a loss. "There seems to be some confusion. There *is* no 1 p.m. flight. The flight was for 1 *a.m.* This *morning.*"

Stunned, I held the receiver away from my head and stared at it. There was a long silence as the news seeped through the fog in my brain and finally sunk in. One *a.m.*? My mind rewound blurred images at full speed to 1 a.m. Where were we at that time? I did the math. At 1 a.m., we were en route from Newark, 35,000 feet over the United States. *How could this have happened?* I screamed in my head, the words pounding my brain. But as the thoughts came into focus, I realized that at 1 a.m., we were in the air because our Newark–LA flight had been delayed. If not for that setback, we would have arrived in LA at midnight, as planned, but there still would have been no way we would have made the LA–El Salvador connection. Our luggage was packed with vacation clothes, and we'd always planned to go home and repack our bags for a funeral. Where had the misunderstanding been? "But . . ." I wanted to argue, to yell at the man on the other end of the phone. I wanted a 1 p.m. flight to miraculously be added to the schedule so we could fly home as fast as possible. I wanted to snap my fingers and be transported 3,000 miles to Mom's side. I wanted to turn back the hands of time. I wanted my brother back . . .

"What about another flight or another airline?" I asked in a desperate whisper.

"There's an American Airlines flight that leaves in forty-five minutes. We could get you on that one." He paused. "If you can make it to the airport by then."

"We won't make it," I whispered. My head was throbbing. "What about a connecting flight . . . or something? I need to get home as soon as possible."

"We could book you via Mexico or Costa Rica. You could leave today, this afternoon, but the problem is the connecting flight will actually arrive in El Salvador later than the next 1 a.m. direct flight. I'm so sorry."

I didn't reply but let out an enormous sigh, a sigh that seemed to express every bit of fatigue and fear and frustration and grief that had churned in me for two days. This felt like a dream, the kind in which I'm trying to run from something but can't get anywhere. As if trying to run underwater.

I'd long ago shed any belief in an almighty, all-loving God. My faith had always been tenuous at best. Raised a Catholic, I was the kid who constantly asked, "But why?" Never satisfied with the answers I was given, I eventually accepted there was just no scientific proof to support any of the beliefs the priests spouted, so the kid I was grew up to be a woman who needed reason, evidence. Yet, despite being a non-believer, today, I cursed the very God I denied. *Fuck you, God! What the fuck do you want from me?! I hate you!* I rubbed my temples.

"Okay," I finally said. "Please get us on the next flight. We need to get home as soon as possible."

"Please be at the airport at 10 p.m. tonight. We will be waiting for you and your husband. And Ms. D'Antonio, I am so sorry for your loss."

With several hours left before the flight, I wandered about the house, stopping to gaze at the collection of family photos that stood on the mantel. Freddie and I had moved in three years earlier, delighted we'd found a contemporary home in move-in condition with high ceilings, decorative moldings, walk-in closets, and arched doorways. We'd fallen in love with the place as soon as we stepped over the threshold. The home had far more square footage than we needed, and we spent a fortune furnishing a bunch of rooms we almost never used, but we compensated for the extravagance by often hosting family and friends for barbeque dinners and festive holidays. Cesare had never made it to the house. We could never fit a visit into his busy schedule.

It was now just past lunchtime on a spectacular Southern California day. Bright and vibrant. Full of life. Birds chirped in the trees, and a brown ground squirrel eyed me from across the yard, flicking its bushy tail. I sat on our concrete deck by the pool and turned my face to the sun. Wrapped in the soft afternoon breeze, I felt safe as tears rolled down my cheeks. The trees swayed, and the rustling leaves whispered with each wave of wind. For a moment, the chaos had quieted.

I thought, *Cesare would have loved it here.*

It was now fifty-two hours since Cesare had died.

I was beyond tired but still hadn't slept. I'd spent what was left of the early morning hours unpacking our vacation clothes, doing laundry, and repacking our bags with clothes appropriate for a funeral.

I was headed to my brother's funeral.

Sometime that afternoon, I logged on to the internet. I needed to know more about what happened *that day*. An engineer by training, I needed facts. Facts, like pieces of a puzzle, must be collected and arranged until a clear picture takes shape and comes into focus. But this fact-finding mission would do more than just recreate a storyline. It was going to ground me and redirect my attention away from my exposed vulnerability. Research and reason and facts—this was how I'd navigated so many natural and human-made disasters throughout my career as an engineer.

In 2001, I'd been part of a five-person unit sent to San Salvador on an American Society of Civil Engineers earthquake reconnaissance team. We'd been assigned to determine the causes and effects of the destruction after two devastating earthquakes hit the area one month apart. The first, which had hit in January and measured 7.6 magnitude on the Richter scale, caused a landslide that wiped out an entire housing development called Las Colinas, The Hills. In a matter of seconds, 200 homes were buried by tons of mud that instantly killed almost 600 residents. Our team arrived at the site after search and rescue operations were already completed and all remains of shattered homes, mangled cars, broken trees, and stripped-away vegetation had been bulldozed and removed. The timing ensured we didn't interfere with rescue operations. We parked our car outside the cordoned area and split up. I was horrified as I toured the site, which was now leveled. An area the size of five football fields was now barren land, everything in its path plowed down. Scanning the span of it and taking in the devastation, I now understood the expression, "looks like a bomb went off." An eerie silence took my breath away as I trekked what I considered hallowed ground.

I stood in the middle of what was now wasteland and studied the deep scar on the face of the mountain ridge where the slide had started. On either side of the ridge, there was heavy vegetation, which indicated the existence of localized perched groundwater that kept the hillside green and lush year-round. The homes that had stood here were most likely the biggest investments in the lives of those who had populated the middle-class neighborhood. The seismic event had occurred on a Saturday just before lunchtime, and I thought about those innocent people who had simply been going about their weekend business, completely unaware that within minutes, their lives would be over. It was almost too much to process.

In my mind, I ran a camera reel recreating the landslide. The combination of the groundwater served as a lubricant, and the low-strength soil that made up the hillside caused a debris flow that traveled approximately 1,500 feet. I imagined the massive flow swirling around where I stood and several hundred feet beyond, destroying everything in its path.

Something on the ground several feet away caught my attention. I bent down and picked up a ribbon badge that read, *It's a Boy!* I guessed it had once been blue—now it was sullied gray by the dried mud. Maybe there had been a baby shower, a hopeful celebration attended by a throng of happy friends and neighbors. Or maybe it had been a party to celebrate the baby's arrival. And now all their lives were over. Or maybe not, I thought with aching hope. Maybe mother and child had been far away that day. Or maybe she'd been home but by some miracle had made it out alive.

Right after arriving in El Salvador, our team heard stories of survivors who shared their firsthand accounts of that day—the randomness of who lived and who died all based on the location of their homes. A housewife who stepped outside to her mailbox had been swept away in the slide debris while her home remained intact, her cigarette still lodged in the ashtray on the coffee table. One survivor was rescued after being trapped for days in the rubble, his body being slowly crushed by mud and concrete, only to die in the hospital. For days, rescuers discovered body parts in the loose soil.

There was no denying the earthquake was a natural disaster, but the scope of the catastrophe due to the landslide was all man-made. Those homes were tucked in a hamlet just feet from a steep, 250-foot-high slope that had been over-vegetated because of a constant source of water. When I'd been a rookie in geotechnical engineering, I'd learned this kind of over-vegetation was likely to lead to tragedy. The earthquake triggered the land movement, but the slope had been doomed to give way sooner or later. The warning signs were there; they'd been there before the homes were constructed, but no one had taken action to protect the people.

For years, El Salvador's government allowed vast deforestation and widespread land development. But the country had adopted no building codes and failed to apply zoning code restrictions for hillside development. Many residential tracts were allowed to be constructed without due diligence to secure safe and stable sites. The victims of the Las Colinas landslide had trusted their government, a complex bureaucratic system, to protect them, but their trust had been grossly misplaced.

It was this investigative mindset that drove me to search for facts about my brother's plane crash. Online information about the accident was scant, and what I could find was just a repeat of what I'd already read on some other site. Everyone reported TACA Flight 390 had departed San Salvador on schedule and under bad weather conditions. Central America had been under siege from Tropical Storm Alma, which was responsible for heavy rain, strong winds, and thick fog. Flight 390's final destination was Miami, with two intermediate stops in Honduras, one in Tegucigalpa, another in San Pedro Sula. There were 135 people on board. The news coverage said the cloud ceiling had been low, causing poor visibility during the approach and that the captain had aborted his initial landing attempt due to heavy fog. Flight 390 circled the airport for approximately half an hour before attempting to land once again. During the second landing attempt, the plane overshot the runway, plowed across a roadway, and crashed onto an embankment before coming to a violent stop in a ravine lodged between two residential communities and a cluster of businesses.

The reports said three people onboard had died, including the pilot and two passengers—one from a heart attack—along with two people who died on the ground when their vehicle was crushed by the plane as it landed on the highway.

I clicked through several sites and found a press conference held by TACA Airlines that day in Tegucigalpa. Roberto Kriete, Chairman of the Board and CEO of TACA Airlines, flanked by high-level executives, confirmed the number of fatalities and expressed heartfelt condolences to the families. By that time, TACA representatives had already informed the next of kin, so he read off their names. He stated that, after all passengers had been evacuated, the injured had been transported to local hospitals and were being cared for. The pilots were the last to be evacuated. Kriete praised Captain D'Antonio for his experience and performance, stating that Cesare's actions in those final moments had prevented the tragedy from being much worse. When responding to questions from reporters, he mentioned "less than optimum braking conditions" as a possible cause for the crash. I was intrigued by TACA's willingness to share this observation despite having very little data at that point to substantiate their hypothesis. It was as if they were privy to inside information. There were so many things that could have contributed to the crash, why focus on that one issue? What did they know? What did "less than optimum braking conditions" mean, exactly? Had the brakes failed? Was it possible there had already been concerns about the safety of the runway? Was there tangible evidence of a runway design flaw? If so, could this tragedy have been easily prevented?

Hearing about the braking conditions gave me a little hope.

Maybe Cesare would *not* be made a scapegoat after all. Hours after the accident, Kriete was already confirming there was at least one conclusion that had nothing to do with pilot error.

I wanted to see what I could learn about Toncontin International Airport's history, so I read about the airport on Wikipedia. Toncontin Airport, located four miles from downtown Tegucigalpa, was surrounded by residential communities and industrial complexes. In 1921, when the airport was just an improvised runway on farmland, a single-engine plane made the site's first landing. Then in 1933, the government of Honduras acquired the land and in 1948 officially inaugurated it Toncontin International Airport. Because the site is surrounded by mountains, arriving pilots have to make a steep ascent to clear the mountain peaks before quickly descending into the valley below. Then they have to land and come to a stop on one of the shortest runways in the world.

I already knew Toncontin had a history of tragic accidents and that because of this, many referred to it as a dangerous airport. After reading a few articles about the crash, I suspected the state of the airport was a hugely relevant factor in the accident. But I wondered if I'd arrived at this preliminary conclusion as a grieving sister. Could I be objective while investigating the crash of a plane my own brother had been flying? Was I falling into a confirmation bias trap? I accepted that I wanted desperately to believe Cesare wasn't responsible for the tragedy. Even Kriete had hypothesized there was at least another possible reason for the crash, but had I latched on to that because I couldn't bear the idea that Cesare could have been negligent, that maybe my brother had caused his own death and the deaths of four others?

I looked away from the screen and thought about what I was asking myself. Yes, I was a grieving sister, but I was also trained as a critical thinker. It was my job to rely on facts, to assess findings with cool clarity and leave emotion out of it. Most of all,

I *wanted* truth, and while I wanted desperately to find that my brother hadn't erred, that he hadn't been in any way responsible for this tragedy, I knew myself well enough to know that self-delusion wasn't something I could tolerate, certainly not in the face of firm, cold facts. I would dive into this investigation, and if the facts I unearthed indicated Cesare had indeed caused this horror, I'd accept that. To do otherwise would be to scoff in the face of evidence, of truth. And I was a seeker of truth. *I* was the girl who had rejected Catholicism because it could offer no evidence. If Cesare had caused all those deaths, those injuries, all that damage, I'd believe it. I'd say it out loud, and I'd figure out how to deal with the emotional toll that would follow. But for now, it was up to me to hunt the truth and follow wherever the evidence led.

I reread the facts of the crash over and over. I examined the online pictures posted by people who had hovered near the rescue site. My eyes scrambled left and right, desperately looking for Cesare, all the while hoping not to see him, terrified of the condition he may have been in. My chest hurt with every inhale and exhale. I tortured myself with every click of the mouse. There was a picture of the copilot being carried away from the wreckage by several men, his white shirt soaked red with blood. He was alive.

There were several pictures of the cockpit taken from various angles, all of them showing silhouettes through the windshield. The reflection on the glass made it impossible to clearly see my brother, but I knew it was him in the captain's chair. Was he still alive at that point? Was he in pain? Was he scared?

I continued to stare at the pictures. Rescue workers had used primitive equipment to pry the cockpit open and release the trapped pilot and copilot. No jaws of life, no sophisticated machines; instead, rescuers appeared to be using metal spears and pickaxes to try to break the cockpit window. It was all so rudimentary. I was enraged at the lack of sophisticated rescue operations that might have otherwise saved Cesare's life. And in every photo, there was someone or something obstructing the view of Cesare's face. I found that so odd. Even in the picture that showed him being rescued from the cockpit, a trail of blood running along his left arm, his face was blocked by a rescue worker.

Rescue worker. Rescue worker. I caught myself thinking about the words we use for people's roles in a disaster. Cesare wasn't rescued. He was being *removed*. When they finally pulled my brother from the cockpit, he was already dead. What good is a euphemism like *rescued*? Just tell me the truth.

Our house phone rang, and I ignored it and wondered how soon the answering machine's memory would fill. Freddie was out picking up our car from the long-term parking lot, and I was grateful he'd left me to scour the internet on my own. I'd been at it for several hours, but my fact-finding mission wasn't even close to being over.

While reading another story about the crash, I saw a picture taken in the morgue. It was of a white body bag, one end of it drenched in blood. The caption identified it as the body of the captain. I let out a gasp as an electric current shot through my chest and a drumbeat thundered in my ears. Right away, I tried to ease my pain with irritation. *Why white body bags?* I wondered as I stared at the offensive photo, my line of sight zeroing in on the sack that contained the remains of my brother. In the US, we use black body bags. Why use white and make the situation more awful for everybody who cares about the person killed?

The red stain that seeped through the fabric left no doubt Cesare had sustained injuries that caused extensive bleeding. I seethed. The white bags were idiotic, and whoever took the picture of a bloodied body bag and launched it onto the internet was a classless human being. An abhorrent violation of privacy and sacredness of death. I could now direct my anger at whoever that soulless heathen was.

Today, everyone with a cell phone can be a photographer or

journalist, recording history through a biased lens, passing judgment, and manipulating public opinion using weaponized images. Thanks to the internet, everybody can play voyeur, judge, and jury. I braced myself for where the Web would land on the matter of Cesare D'Antonio—hero or villain?

I heard Freddie pull our car into the driveway. Then he walked into the room. "I'm home," he said, reaching for me and wrapping me in a strong hug. "We should get ready for the airport."

I shut down the computer and headed upstairs.

5

All Aboard

Sixty-two hours after Cesare's death, Freddie and I arrived back at LAX. It was 10 p.m., and the drive to the airport had been a quiet one, nothing like the chattering, excited ride we'd had to the airport just two weeks earlier at the start of our Spanish adventure. Tonight, even the air in our car seemed to be grieving. Having learned our lesson about airport transportation, we parked our car at the Parking Spot on Century Boulevard, two minutes from the airport proper.

As we walked through the airport's sliding doors, a young man near the counter raised his hand to get our attention. He rushed over to us. "Ms. D'Antonio? Please come with me." TACA Airlines had been waiting for us. I wondered how their representatives had recognized us amongst the rest of the crowd. Was it obvious we were part of a grieving family? What gave us away, our dark clothes or the grief stamped on our faces?

We reached the TACA Airlines counter, and I introduced Freddie and myself softly, not wanting to call attention to us but knowing the attention was inevitable. "We're the family of Captain Cesare D'Antonio," I said, handing over our passports.

With warmth in his dark, brown eyes, the young man behind the counter reached for the passports. "Of course, good evening." Then he lowered his eyes, and his fingers began to dance over the keyboard as he clicked our information into the system. The light reflecting off his computer screen cast a strange glow on his somber expression and made me think of Van Gogh's *The Starry Night*, with its swirls of blue and yellow. So strange, the things that come to mind when the soul is in shock.

The young man motioned to his supervisor, a young lady in a business suit, who walked over, smiling politely. Standing side by side, they had a brief exchange in hushed tones as they concentrated on the screen. After attaching a priority label to our bag, the man placed our luggage on the conveyor belt while we answered all the security questions like robots.

"Yes, we packed our bags ourselves."

"Yes, our bags have been with us the whole time."

"No, no flammables."

The airport was alive with activity. Passengers navigated the switchback queue with their luggage carts. A baby girl strapped into a carrier on her father's chest wailed a shrill, piercing howl. A janitor mopped up near the restrooms. I stood watching the scenes as if I were a spectator in a slow-moving picture show. After a short time, the young man with the warm, brown eyes handed us our boarding passes and pointed down the hall. At this point, I'd slipped into some kind of safety bubble and had tuned out all the airport's noise and bustle. The man's lips moved, and I made out the word "security." I nodded.

As Freddie and I turned to move on to the next step in the painfully long process of going home, the nice man and his supervisor reached out, gently touching my hand as I clutched the boarding passes. It was this human contact, their warmth, that jarred me out of my dream state and back to the activity of the airport. It was as if a radio had been turned on full blast.

"We're so sorry for your loss," they said.

"Thank you," was all I could mutter.

Freddie and I sat shoulder to shoulder at our crowded gate. My head was pounding. I still hadn't slept or eaten since getting the news, and I rubbed my hands together, as if that would bring some warmth to my body that now felt so cold. I glanced at my watch. We'd be boarding soon.

A thin young lady with wire-rimmed glasses sitting one seat over turned to me and smiled. I forced a smile back and wondered if she was flying solo. Maybe she was trying to make contact to ask if I'd watch her bag while she ran to the restroom. She leaned over and asked, "Did you hear about TACA's plane crash?"

I stared at her, stunned, then sat paralyzed for a moment while electricity shot through my body. She looked confused during the awkward silence, and I rose and walked away.

Oh my god! I thought. Afraid I'd break down in public, I rushed to the nearest restroom. Taking refuge in a stall, I allowed the minutes to tick away until I felt ready to face the world again.

Our flight was announced on the intercom, and I rushed back to Freddie who now stood in a far-off corner, away from the crowd. He pulled me into one of his hugs that always made me feel nothing could ever hurt me. *In good times and in bad.* I realized this was our first major crisis since saying "I do." Three miraculously fortunate years with no crisis whatsoever, small or big. How lucky we had been. We remained wrapped around each other until our row was called, then we headed to the front of the line where an airline attendant collected our boarding passes, scanned our passports, and smiled sympathetically. I thought, *He knows.*

"Have a good flight," he said, returning our documents.

"Thank you," Freddie said as he squeezed my hand.

We turned the corner toward the jetway and came face-to-face with the enormous nose of the majestic plane that would take us home. The great white bird sat regally on the tarmac and seemed to stare back at me through the terminal's floor-to-ceiling windows. With outstretched wings and sleek lines, it stood at attention, looking elegant, highlighted by the beam of the full moon on this dark night. I was grateful this beautiful emissary stood ready to reunite me with my heartbroken family.

As we stood against the glass wall, I had a direct view of the tiny cockpit, fully lit and clearly visible through the wide-open eyes of the jet. I had to blink at what I saw, what I *thought* I saw? There, right before me, was my brother. Unexplainable, and yet there he was.

"Cesare," I whispered under my breath. I stood frozen, trying to focus on the man in the captain's seat, at Cesare. His forehead was pinched, as if he was concentrating on something. I could see he still sported his trim beard. Cesare sat upright in the captain's chair, turned his head slightly, and I caught his profile. Years earlier, he'd stopped wearing his diamond earring because airline regulations prohibited it. I saw him extend his arm to grab something to his right, and the epaulet with the four golden bars on his shoulder flashed light in my direction. Cesare seemed to be frowning as he said something to his copilot. Then he turned and stared back at me through his Prada sunglasses. I remembered when he bought those glasses—it was during a vacation he took to Rome. He'd come back so proud of them. After that trip, he almost never took them off. Cesare smiled at me, that amazing toothy smile! I could hardly breathe.

Suddenly, a voice making announcements boomed over the intercom, and the spell was broken.

Freddie gently tugged my sleeve. "We should get going."

With my head lowered, I shuffled down the jet bridge, and as I stepped into the plane, I looked left, where my flight's pilot and copilot sat and busied themselves with the details of flight prep. There were no sunglasses. There was no Cesare.

I stared at them, hypnotized by how small a cockpit was on a jet this large. This was how small Cesare's cockpit must have been. The impact had been dead-on and had crushed the space where he sat. There was no way Cesare could have survived the crash.

A petite, dark-haired flight attendant greeted us with a warm, friendly smile and motioned us to seats 1C and 1D. There was no pity in her eyes, so I assumed she didn't know.

I took the window seat, and we made ourselves comfortable. Before long we were taxiing, almost ready for takeoff. The crew demonstrated the emergency procedures, but nobody paid attention as the flight attendant stood in the aisleway holding a short section of seat belt, clicking it closed and unbuckling it again.

Buckled.

Unbuckled.

Without looking down, I fingered my buckled seat belt. There was talking all around as passengers ignored the safety demonstration. I wondered if the same apathy had been on display before Cesare's last flight. Maybe those who survived the crash were the ones who had paid attention that day. I made a mental note to try to find out.

The flight attendant was now pointing to the emergency doors, two in the front of the plane, two in the midsection, and two in the back. I strained to hear all the safety instructions, irritated that nobody else seemed to care, that we'd all grown so complacent no one paid attention anymore. There was a smugness in those of us who considered ourselves pro-travelers, as if we'd always known what to do in an emergency. But for most of us, that premise had never been tested. In the seats right around us to several rows back, I heard people chattering. I winced. Even here on a plane owned by the airline that had suffered a major airplane crash just two days earlier, a crash that had left several dead and many more injured, it seemed few passengers cared about safety instructions.

"Should the cabin become depressurized," said the voice on the intercom, "oxygen masks will drop down." The pretty flight attendant pulled on the oxygen mask, stretched the elastic bands, and pretended to place the yellow silicone facial cup over her mouth.

I reached for the safety pamphlet in front of me as the flight attendant encouraged us to read the emergency information. The laminated trifold card indicated we were in an Airbus 320, just like the one Cesare had flown three days earlier. We were sitting in first class just behind the forward galley and bathroom that separated the cockpit from the passengers. I closed my eyes, trying to stop my brain from recreating Flight 390. *Stop*, I urged myself.

But it was no use. At the moment the plane started to rumble down the runway, I imagined those final twenty-six seconds of Flight 390 like a clip of a feature film playing over and over. Our plane accelerated, and I gasped softly, then squeezed Freddie's hand. He squeezed back.

Our plane was now screaming down the runway—or was that me screaming? The aircraft's wings wavered and bounced as we gained speed, and the doors of the overhead compartments rattled and clicked. I could feel my pulse throbbing in my neck. Out the window, brightly lit buildings seemed to race by in a violent blur. Pressed back into my seat, I clutched the handrests with white-knuckled hands. And then we were airborne. I held my breath. The drone of the aircraft's engine was all I heard.

The plane didn't drop from the sky. It continued to gain altitude, and I felt myself pressed even harder against my seat. Then I heard a grinding noise beneath us and felt the landing gear retreat. My heart continued to pound, but I peered out the window consciously breathing in and out. Breathe in and out. In and out. In and out. Then the plane began to level off, and the volume of the engines dropped a notch. I continued to stare out at the amazing expanse of lights below, and somehow that soothed me. My beautiful Los Angeles. City of Angels.

It was now almost 2 a.m., and the lights in the cabin were turned off. All around me, passengers tipped their heads onto the travel pillows that curved around their necks. I sat listening to the hum of the plane, a modern-day lullaby. A plane's white noise had always compelled me to drift off to sleep, but on this night, I sat wide awake.

During Cesare's fifteen years at the airline, I had flown with him twice. Only twice. Once with Freddie from LA to El Salvador for the Christmas holidays. Cesare had made sure we were upgraded to first class. On that night, once we reached cruising altitude, he came by to check on us. He crouched down to eye level and flashed his crooked smile. "Hey, you guys. Okay? Are you being taken care of?"

"Of course," I replied, raising my wine glass. I pointed to the cockpit. "Shouldn't you be in there flying this beast?"

"This thing practically flies itself!"

"Well, why don't you go back in there anyway and do your job? Thirty-five thousand feet is a long way to fall!"

Cesare laughed. "You're lucky to fly with the finest."

"The finest what?"

It had always been like that with us. So much good-natured sniping and ribbing. Even the subject of airline disasters wasn't off-limits. That's just how comfortable we always were.

When Cesare had just started out with the airline as a First Officer, I'd flown with him on a commercial flight during a Christmas holiday. He'd walked to the back of the plane where I was sitting.

"Hey, you wanna come and take a look at my office?" he asked, pointing to the front of the plane. "I'll give you a tour of the flight deck."

"Sure!" I responded, suddenly feeling like a ten-year-old kid again.

The expression "cabin fever" gained a whole new meaning as I stepped into the cockpit, surprised to see how small it was.

Cesare, all six feet of him, bent forward slightly to avoid bumping his head. At five feet, five inches, I didn't have the same problem. He introduced me to the captain, who turned to look at me and flashed a wide, welcoming smile that accentuated his bushy moustache. I had a great view out of the front window at the full butterscotch moon that illuminated the night sky, the type of moon that always reminded me of the expression "love you to the moon and back."

"We're flying in a Boeing 737, which uses yokes to steer the plane," Cesare said, pointing to the U-shaped handles the captain was holding. The instrument panel was aglow with multicolored lights and gauges. There were knobs, dials, and controls everywhere I looked, and a rainbow of colors streamed above my head from the overhead panels. I stood just behind the two pilot chairs, somewhat daunted and intimidated, afraid to touch anything for fear of accidentally sending the massive plane into a tailspin.

Finally offered the jump seat, I sat and watched my brother talk on the radio. The captain provided some instructions to the copilot. A switch to autopilot.

That day it became so clear to me how much Cesare loved flying, and my heart swelled at the sight of him, the sound of him so competent, doing what he loved, and clearly feeling sure of himself.

After a while, Cesare, my little brother the First Officer, walked me back to my seat. I wanted to yell out to everybody on board, "Hey everyone, this is my kid brother who's helping fly this massive bird and will get you to your destination safe and sound!" He kissed me on the cheek, and I sat beaming while everyone around me watched.

The memories sent streams of tears down my cheeks, and I was grateful for the darkness of the cabin. Then, as if on cue, the pretty flight attendant appeared from behind the separator curtains.

"Can I get you anything?" she asked. Caught off guard, I couldn't hide my tears, and her eyes filled with pity.

"No," I said, "I'm fine."

She walked away and returned shortly with a wad of napkins and placed them on the blue blanket covering my legs. "If there's anything you need, anything at all, please don't hesitate to call me."

I thanked her. *She knew.*

A few hours later breakfast was served, but I couldn't bear the thought of eating. I still hadn't eaten anything since receiving the news. I told the flight attendant I'd have some coffee. She leaned over and said softly, "You need to eat something. How about some fresh fruit?" I was touched by her kindness. I smiled and nodded. She placed the little bowl of kiwi slices and berries on the pullout tray, and I picked at it. I was glad to see Freddie eating his scrambled eggs. I needed him to remain strong for me.

A few minutes later, the captain's voice came through the overhead speakers. "We will be landing in San Salvador shortly."

Everything was in order, breakfast dishes collected, seat belts fastened, and backrests upright. Just like Flight 390 only hours before.

I gripped the armrests. What if Tropical Storm Alma didn't give us a break? The jumbo jet descended ever so slowly as it sliced through thick cloud cover, then emerged long enough for me to catch a glimpse of the next dark patch ahead. The tiny windows were lined with a thick layer of condensation, making the inside of the cabin dark and gloomy. My stomach was queasy as the jet rattled and dipped with each air pocket. It wasn't long before the wheels met the runway with a sharp jolt. I squeezed Freddie's hand and saw tears rolling down his face. I wondered if he too was thinking of Cesare and if he shared my anxiety. This was the stage of TACA 390 when it had all gone bad. Up until the plane had touched down on the runway, that flight had been just another air travel experience for everyone on board.

I watched the wing flaps lower and focused on the landmarks as they zoomed by—just as the passengers had done a couple days earlier. Everything had seemed normal, but it wasn't.

Our plane continued down the runway slower and slower and slower yet, and then it finally came to a stop. On solid ground. Safe and sound. I finally let my breath out. How long had I been holding it?

The ground was wet, but it wasn't raining. There was even a tinge of sunlight seeping out from behind the clouds. I looked out the tiny window at the sky. It appeared Alma had cut us a break after all.

Feeling my brother's presence, I whispered, "Thanks, Cesare." And I wondered if the roles had changed and it was now his turn to protect me.

The cabin bustled with passengers talking and collecting their belongings. Compared to the excited travelers all around me, I seemed to be moving in slow motion. Freddie grabbed my hand, and we waited for the doors to open, only moments away from stepping out to a reality I'd been dreading since receiving the news. It was 6:30 a.m. I put my sunglasses on.

The flight attendant leaned over once again. "Excuse me. Are you related to Captain D'Antonio?"

"He's my brother," I answered, well aware I'd used the present tense.

"I am *so* sorry," she whispered, placing her hands over her heart. "I knew him. He was such a wonderful person."

"Thank you," I said. "I'm sorry, I didn't get your name."

She told me her name, and right away I forgot it. In the days immediately after the tragedy, we received much kindness from strangers, and they provided huge comfort. I remember their faces, and I carry them in my heart for their generous compassion.

Placing my hand over my heart, I said, "I thank you so much for taking such good care of us tonight."

The doors opened, and Freddie and I stepped off the plane.

6

Saying Goodbye

As we stepped off the plane and onto the jetway, an airline representative greeted us. I suppose it wasn't hard to spot us; we were the first ones off the plane, dressed in black. We were headed straight to the funeral home. Our escort introduced himself, then offered his condolences and made general pleasantries. "Good flight?" "This way to customs." "How many pieces of luggage?" The humidity in the air made it hard to breathe, and I found myself responding in monosyllables.

The trek through the crowded corridors and down the escalator was a blur. We breezed through customs and immigration and were reunited with our luggage, discreetly set aside, although our flight's assigned carousel hadn't yet started turning. The escort ushered us out of the building where a driver in a black SUV waited for us.

Time seemed warped. Scenes sped up as if on fast-forward, only to slow back to normal speed. Perhaps it was exhaustion that played tricks on me. A lady stepped out of the SUV and embraced us.

"My deepest condolences. Captain D'Antonio was a wonderful human being," the woman said. She was also dressed all in black. I was tired and still in a daze, but I was touched by her kindness. "It's a pleasure to meet you," she said. "My name is Marta Liliam Martinez, head of TACA's Human Resources Department. Call me Marta Liliam."

Rain began to fall, so we quickly loaded our luggage into the SUV and said goodbye to our escort.

When we were settled in the vehicle, I said, "I was told Cesare is at Montelena Funeral Home. I hope you know where that is because I don't." It was this kind of lack of control that made me feel lost and out of balance. But Marta Liliam had been in contact with my family since the tragedy.

"Don't worry," she said, placing her hand on mine, "I know where it is." Her voice was comforting, and I immediately felt at ease even though I was on my way to be reunited with my dead brother and my grieving family.

"How is the first officer doing?" I asked.

"First Officer Juan Artero is in bad shape and has been flown to a hospital in Miami. His injuries are serious, but he's expected to survive. He has multiple broken bones and will require reconstructive surgery, but they tell us he'll be okay."

I nodded, "I'm glad to hear that." I didn't know him but had read in one of many articles that he was only twenty-six years old, just starting out. I'd thought about him many times in the past two days and deeply hoped he'd make a full recovery.

The rain continued to come down in sheets.

On the drive from the airport to the funeral home, we weaved in and out of several quiet neighborhoods blurred by the hammering rain that was intermittently swept away by the wipers. I didn't recognize any of the homes or businesses. Had it been that long since I'd been here, or was my mind clouded with grief and exhaustion?

Before long, Marta Liliam steered the SUV into a large complex with a big stone waterfall and *Montelena* spelled out in big brass letters. We passed through ornamental gates, then entered a beautiful compound sprawling with full serene gardens of begonias, pansies, and marigolds. In every direction was a blanket of green. Then we entered a parking lot made up of brick pavers and gravel, and as the tires crunched over the gravel, my heart rate sped up. We were close now. We'd be inside soon. Marta Liliam parked the car and said she'd have someone take care of the luggage. I got out of the SUV, stared at the building, took a deep breath, clutched Freddie's hand, and headed toward the double doors. It was now almost seventy-two hours after Cesare's death.

We entered a big open room furnished with large black leather chairs and sofas. The walls were all glass, and from my vantage point, I could see heavy sheets of cascading rain outside. I hated the rain. So had Cesare. I looked away from the windows and there, serenaded by the thundering downpour, in the center of the room was my mother, sobbing uncontrollably. As I approached her, Mom caught sight of me, her face contorted with pain. Her legs buckled as I wrapped my arms around her, and I held her frail, trembling body firmly. She was weeping like a child, without shame, without self-consciousness. The sight and sound of my mother's agony was more than I could bear. Mom had always been the pillar of our family, but as I held her tightly, she was a diminutive figure lost in my arms.

"Cesare's gone," she repeated between sobs. "Cesare's gone." I said nothing. We stood like that for a very long time, each gaining strength from the other. After a few minutes she let me go, and Freddie, now sobbing himself, took her in his arms. I reached for my sister, her fair skin flushed, her eyes red and puffy. Mars held me tight and whimpered into my shoulder.

Mars was the one Mom leaned on during the first few days. As the sensitive daughter, she was well-equipped to offer the emotional support Mom needed, something that doesn't come naturally to me. I couldn't think of anyone better to serve as Mom's rock. But there had been logistical matters to deal with as well, matters that would have normally fallen to me to handle. Despite her overwhelming grief, Mars had navigated the gauntlet of decisions to be made and actions to be taken. I was proud of her.

I'd read somewhere about the US military's dignified transfer of remains for those killed in the line of duty, and I was impressed with the level of respect and care taken during that solemn journey. Protocols included administering the appropriate amount of formaldehyde during embalming to allow for the possibility of several days of travel, proper dressage for family viewing, and assigning an escort for delivery to the next of kin. While certainly not the same circumstances, I'd expected the transport of Cesare's remains would have been handled with a similar degree of respect. But no such care had been taken when TACA returned my brother to us. Mars was one of the first to see Cesare in the casket upon his arrival in El Salvador.

Mars pulled me aside and whispered, "Cesare arrived from Honduras dressed in a simple short-sleeved white shirt with a red tie and black slacks. He wasn't even dressed in his captain's uniform. His body lay crooked in the coffin." As she spoke, she lifted her shoulder blade and swayed her hip in the same direction. "Crooked like this," she said. "It was as if he had been carelessly laid down or maybe he had been jostled on the flight." I stared at her in disbelief. "But the worst part was his head. It was so carelessly bandaged. The dressing started at his forehead and was wrapped so it ended in an unnatural point at the top of his head. It made him look like a conehead."

What? Who had prepared him? I was livid that whoever had been assigned this task didn't have the standards or grace to provide Cesare the honor and dignity he deserved. I was furious for Cesare, and I was furious for the rest of us.

"I just knew Mom couldn't see him like that," Mars said.

Mars explained she'd arranged for Galo, Cesare's best friend, to drive to Cesare's condo and retrieve his captain's uniform so he could be dressed properly for his final trip. I was heavy with guilt that I hadn't been around during the hell of the past few days. I was usually the one taking control, solving problems. But I was here now, and I was going to do everything I could to make up for the lost time. I knew the days ahead would be full of logistical decisions to be made regarding Cesare's affairs. I was here now and would shoulder those myself.

In a low voice, Mars whispered, "They just took Cesare away. He needed more embalming. The Hondurans screwed up, and his body is starting to break down."

I stared at her, my eyes wide open. I must have misunderstood. They screwed up. *They screwed up?* How does that happen? The patient is dead. You drain their fluids and pump all cavities with chemicals. What level of ineptitude does it take to screw that up? It was like some obscene joke because Cesare had no tolerance for incompetence. Now his corpse had been subjected to unthinkable ineptitude.

"Are you fucking kidding me?" I said.

Mom walked over from where she'd been sitting on the leather couch and wrapped her arms around me, as if trying to quell my anger. Her strong embrace, the powerful human connection, jarred me back to the moment and calmed my temper. Despite her unimaginable pain, she was trying to comfort *me*. As the matriarch, Mom had always set the tone and direction for our family. It was clear this was no time for anger as she clung to our remaining family.

Except for the sound of rain beating on the ground outside, all was quiet inside the funeral home. I sat next to Mom on one of the oversized leather sofas with my arm wrapped around her shoulders as she sat quietly fidgeting with a wad of tissue that slowly disintegrated in her hands. The silence was suddenly interrupted when we heard a shuffling noise behind us. Four men dressed in black suits wheeled in a silver-colored coffin. Cesare's coffin.

Mom sprang from the couch, then wobbled to regain her balance. "I want to see him."

"Mom, wait!" I said, trying to steady her.

She started to cry again, and she followed the casket into the room, the rest of us trailing behind her. Before long, our little family gathered around her next to Cesare's coffin, and one of the attendants lifted the lid until it came to rest in an upright position. There he lay.

I hadn't tried to delude myself that my brother was really, definitely dead, but seeing him now in the box, I felt what it meant that his death was no longer an abstract reality. It was now a hard, tangible truth.

Standing there, I had the feeling the floor had collapsed beneath me and as if in a carnival house I was free-falling in an uncontrolled spiral into a deep and dark nothingness. Tumbling head over heels. Falling and falling and falling deeper yet. The walls around me appeared to be melting, the room caving in on me.

Cesare lay in the casket in his full captain's uniform. He was still and beautiful, and I saw no signs of trauma.

"There's a bruise on his right cheek," Mom whispered. But I couldn't see it. What I saw was my handsome brother, his trim beard nicely groomed, his expression as peaceful and perfect as if he was taking a nap on a summer afternoon. His face was a little pasty, perhaps from the excess makeup the mortuary used. I guessed that if I were to run my fingers along his cheek, I'd feel the roughness of heavy rouge caked on. I wondered if this was a normal amount of makeup, or if he required extra layering to cover his wounds.

I had seldom seen him in his full uniform, but there he was dressed in his jacket with the four gold stripes on the sleeve cuffs and his captain's hat, its visor discreetly lowered just a bit more than normal to cover the bandages on his head. His left hand lay over his right. I counted all ten fingers and thought of how people count the fingers and toes on newborn babies. How odd, I thought, that here I was doing the same as we got ready to bid Cesare farewell.

Cesare was now in eternal slumber, his dancing eyes forever shielded. There was no sign of the loud, raucous laughter that had announced his presence whenever he was in a room, the big, happy sound that always made him the center of attention. Forever gone was that bright smile with the slight chip to his left front tooth. I never asked him how it got there. Why didn't I ever ask him?

It was now midday, and I could hear hushed whispers behind me. People dressed in black were streaming into the room. Mourners embracing, touching, comforting, weeping.

"You're finally here," someone said to me. *Clearly*, I thought.

"Words fail me," said another. I certainly understood that.

"Your mom can now lean on both you and Mars," said someone else. The words stung.

I told the story of our long, convoluted journey home to anyone who asked because it let me focus on something besides my brother being in a box. I stayed in control by being a graceful hostess, an ambassador at an official farewell gathering. Greeting each mourner, I mingled and made sure everyone was taken care of, engaging in brief snippets of conversation before moving on to the next person. "Would you like some coffee?" "We have bottled water." "Please have a seat here on the sofa."

A tall, distinguished man in a crisp pin-striped business suit introduced himself as Roberto Kriete, Chairman of the Board and CEO of TACA Airlines. "I just want you to know that during those final moments of the landing, Cesare was a consummate professional," he said, touching my forearm gently. "Under peak stress levels and the direst of circumstances, Cesare maintained his composure and focused on what needed to be done. When he realized he was in trouble, he remained calm and had the wherewithal to shut the engines off to prevent an explosion upon impact. The plane was fully fueled for a flight that was in its initial leg and would have been a Molotov cocktail if not for his heroic actions." He spoke slowly and deliberately, enunciating every word as if wanting to ensure I grasped its full meaning. "Cesare was responsible for saving the lives of 132 passengers. Cesare was a hero."

A *hero*. My heart swelled.

Cesare had shut the engines off? I made a mental note to verify this claim after the official accident investigation concluded.

A young petite lady with long black hair sat crying in a corner. I excused myself from Roberto Kriete and walked over to sit next to her. She introduced herself as Cesare's manicurist. I thought, *My brother had a manicurist? I don't even have a manicurist.* What else didn't I know about him?

"I just saw him a few days ago at his weekly appointment," she said between sobs. *Weekly* appointment?

I smiled and thanked her for taking such good care of him, thinking he'd be happy knowing his nails were nicely groomed for his final journey.

I stood up and walked over to where Cesare's best friend, fellow TACA captain, Galo Conde, stood. He said, "Hey, how are you holding up?" Then he wrapped an arm around me and handed me a bottle of water. "Water will have to do for now. We'll get you a drink later." I smiled but didn't respond. Being this close to Cesare's best friend brought up memories of their wild adventures and carefree days. I could feel my face pinch as I was about to cry, and he flashed a mischievous grin. "Hey, do you know there are several former girlfriends here? Hopefully, we don't have a catfight break out." He tipped his head toward a woman who was just walking in. "That one has a temper on her." I laughed and dabbed my eyes, grateful for his brotherly banter.

I glanced around and saw quite a few young women had arrived in the past couple of hours. There were tall and petite women, dark-haired and fair-skinned women. Some young and some not so young. Our family had met a few of Cesare's love interests, but some of these others we were meeting for the first time. I scanned the women and concluded Cesare didn't have a physical type. They were all physically beautiful, but there was also something about them all that convinced me their beauty went beyond that. Perhaps it was their confident yet unassuming demeanors, heads held high as they walked the crowded room and introduced themselves to strangers, joined conversations, shared stories.

I wandered about the room trying to stay in motion, and everywhere I turned, someone wanted my attention. Several of Cesare's coworkers stopped me to tell me about Cesare's professionalism and dedication as a pilot. One young man in first officer uniform whispered, "Captain D'Antonio was my mentor." His voice cracked, and he took a gulp of his bottled water. "He took me under his wing and loved to share his knowledge and experience with me. He was never selfish that way, you know?" I nodded.

The room was packed with people, and I tried to greet every one of them and listen to their stories. I found comfort in knowing my brother had touched so many people's lives. And I realized there was so much about him I didn't know. There were friends, colleagues, coworkers, and love interests I'd never met—this room was full of them, and I was hungry to fill this void with all the anecdotes they could offer me.

Later that day, we held a service for Cesare in the adjacent chapel. Our little family trailed behind as the pallbearers rolled Cesare's casket toward the chapel entrance. Lining the long walkway toward the altar stood the honor guard made up of alternating pilots and flight attendants in full airline uniform. I was touched and proud that Cesare warranted this ritual of respect and solidarity by his colleagues who had so generously kept our family company during the wake and showered us with their kind words. They had come out en masse as a show of unity and support. Their presence was a tribute to Cesare, their fallen comrade, who had died in the line of duty.

I wondered what they were all thinking right then. I wondered if it occurred to them how easily it could have been any one of them that day.

A somber funeral hymn filled the church, and I speculated about which artists Cesare would have included in a playlist for his service. Maybe a little Frank Sinatra. Aerosmith. Luis Miguel. Barry Manilow. Vicente Fernandez. He had eclectic musical taste. As the casket made its way down the aisle, each pilot removed his hat and placed it over his heart, and the flight attendants bowed their heads. I was deeply moved by this traditional gesture of an honor guard, and I knew wherever Cesare was, he was proud. From behind my sunglasses, I saw the ashen faces of those standing guard, many who were crying.

By the time my sister, mother, Freddie, and I reached the altar, my body felt heavy, and I had trouble catching my breath. Adrenaline had kept me in motion for three days, but now I was spent. We all sat in the first row, my sister and me on either side of Mom, Freddie next to me holding my hand. The priest at the altar was dressed in violet vestments, the color meant to symbolize mourning, and I knew there were other layers of clothing underneath the outer garment. It was all part of the pomp of Christianity. I sat there thinking all those layers of dressage were symbolic of the need to cover up the Church's misdeeds.

The service got underway, but I was only half listening to the priest's chatter. "Body of Christ," droned the man at the altar in the purple garments. Just as it had been when I was a girl, forced to attend a Catholic school, made to endure tedious lectures and sermons and rituals, I tuned him out. My mind traveled back in time to elementary school.

Blessed Sacrament School. A coed institution where my Catholicism was founded. Father Mark Falvey, a Jesuit priest, was an assistant pastor. He was a kind old man who walked the school campus in traditional black cassock and collar, always greeting us kids with a big smile. When the playground was abuzz with squealing girls in red plaid jumpers and rambunctious boys in blue pants and white short-sleeved shirts, Father Falvey roamed around and made sure we were safe and played fair. During dodgeball games, when the ball wandered out of bounds, he was there to throw it back our way. He always seemed jolly, with that protruding belly, rosy cheeks, and the big smile—always that huge smile.

Our school protocol dictated mass every two weeks, and quite often Father Falvey officiated the service. One Friday before mass, my classmates and I had attended confession so we could receive Holy Communion. I always found it uncomfortable to enter the dark little closet-size room where I'd have to sit and wait for a priest to slide the screen door open and then judge me. It was a ritual founded on an imbalance of power and shrouded in secrecy.

"Bless me, Father, for I have sinned. It's been two weeks since my last confession," I recited the words I'd been taught to say.

"What are your sins, my child?" the silhouette behind the screen asked.

I rattled off the sins of a ten-year-old girl. "I lied to my mother three times, I fought with my brother seven times, and I took extra lunch money."

"I absolve you of your sins. Pray an act of contrition and two Hail Mary's," said Father Falvey, sternly.

I found the whole ritual to be somewhat of a charade. Often feeling guilty that I hadn't committed enough sins or sins worthy of a priest's time, I'd fabricate a few. I wondered if I was the only one not entirely honest in the little dark confessional. And I wondered what might happen to me for lying.

Having rattled off my penance, there I stood in line, free from sin, waiting for Father Falvey to grant me Holy Communion, perhaps the most important sacrament in Catholicism. When I arrived at the front of the line, he reached into the gold chalice in his hand and pulled out a round white wafer, then held it in the air and said, "Body of Christ."

I lifted my right hand to cross myself as we were all taught to do, and my hand bumped the goblet and knocked the wafers onto the floor. My face felt on fire as all eyes lasered on me. The scene played as if in slow motion, including Father Falvey's startled expression as he struggled to hold on to the cup while the wafers flew out and onto the floor. I dashed out of the church, riddled with embarrassment and guilt.

Clearly, Father Falvey knew I would feel awful, and he tracked me down. In his usual kind way, he placed his hand on my shoulder and said, "It was an accident. You did nothing wrong." I couldn't have loved that man any more than I did at that moment.

At the end of my fifth grade, Father Falvey died, and I walked with the rest of my uniformed classmates in single file toward the church altar where his coffin awaited. Father Falvey lay in the brown casket in the dimly lit church wearing a white cassock dress with a purple sash over his big belly. Around his neck hung a heavy silver chain that held a large pectoral cross with the twisted body of Jesus nailed to it. I stared at him and wondered how heavy the cross would have felt on Father Falvey's chest.

Father Falvey was gone. But he'd taught me to have faith, faith in good men. Faith in the Catholic Church. Faith in authority figures.

But in 2007, I learned Father Falvey had done the unthinkable. I found out he was the red-horned demon himself. The *Los Angeles Times* reported that for years this "man of God" had sexually abused four girls and five boys from my school, right up until 1975 when he died. Falvey had encouraged kids to fish candies from his pockets, very close to his genitals. According to the *Times*, a fellow priest walked into a classroom where Falvey was in the midst of molesting a child. When the other priest discovered Falvey's crime in progress, his reaction was to scold Falvey for not locking the door.

Countless children and families were forever scarred by that devil in a robe. And by failing to hold its own priest to the laws of its so-called God, the Catholic Church was complicit in yet another sex crime. One of the girls Falvey molested tried to kill herself. Had she succeeded, the Church would have condemned the act as a mortal sin. They would have denounced and shamed the victim while shielding the aggressor.

I was well aware of the Church's failure at a global level. The cover-ups, the erosion of trust. But Father Falvey was someone I'd known. Someone I had trusted. A childhood hero who had fallen so far from grace.

My parents had carefully selected the Blessed Sacrament School to establish their children's religious foundation. They expected the institution would help strengthen our values and further guide our moral compass. And during my years as a student there, I felt safe. I believed in the institution. I trusted. Now I couldn't process what it meant that during all those years, unbeknownst to me, I'd been in danger, walking a razor's edge between good and evil. I wondered if I knew any of the victims. And had there been others, beyond those the newspaper article referred to? Of course, there had been. Fear and shame keep countless victims of sex crimes quiet.

Why had I been spared? What had kept *me* safe? Whatever it was, it sure as hell wasn't the Catholic Church.

Sitting in a front pew at my brother's funeral, I looked up at the wooden figure on the cross that everyone around me seemed to revere. Centuries of church followers had prayed to this man. A *man*! A man who hadn't even been able to save himself from a cruel, savage death, and yet, the system proclaimed that he'd save us all. Save us from what? Death, destruction, abuse, war, disease? If this was the Catholic plan, they needed a better one because this almighty God was failing miserably. A God who was supposedly all-loving, all-knowing, and all-powerful would certainly not allow newborn babies to be born infected with AIDS or allow terrorists to destroy skyscrapers and kill thousands of innocent people. He wouldn't allow sexually sick priests to prey on children. He wouldn't allow famine, the Holocaust, or plane crashes because of short runways and shitty weather.

"Go in peace," said the priest, signaling the end of the mass.

I stood up to leave the church and thought, *Go to Hell.*

We turned toward the doors following Cesare's coffin back through the line of ceremonial guards who stood erect, this time, each of them holding a single yellow rose. Cesare had been a member of this system, TACA's human infrastructure. Men and women who took to the skies with the noble duty to safely transport the public from one part of the world to another. And on land they leaned on each other with a sense of loyalty that now extended to include our little family. I, having always assumed the role of protector, suddenly felt protected by this extended family. It was as if they transmitted their collective energy and held us in their caring hands. Respect. Competence. Professionalism. Humanity.

As my family made our way down the aisle, Cesare's colleagues honored him by bowing their heads, kissing the yellow rose they each held, and placing it on the casket. By the time we reached the doors, the coffin was covered in a mound of yellow roses, their bright petals emanating a brilliant glow.

I walked the one hundred feet or so from the chapel to the mortuary and sat alone in one of the funeral home's large black leather chairs. I'd now gone more than three days without sleep, and my body was heavy with exhaustion. Before long, Irene, one of my best friends, sat next to me on the chair's armrest and placed a hand on my shoulder. "Have you thought about who's going to give the eulogy? Would you like to say a few words about your brother?"

I stared blankly at her. "I don't know. I hadn't thought about it."

"Do you want to?"

I was quiet for a while. I really hadn't thought about it.

When death marches in slowly, there's plenty of time to think about such things. Loved ones have time to designate someone to give the eulogy, to dwell on preparing a perfect narrative, to design beautiful memorials. But when death comes crashing in unannounced, a family reels and stumbles through the details.

I sat there wringing my hands. But I knew I'd deliver the eulogy. It was my duty. The burial would be in a couple of hours, and I hadn't prepared anything, nor did I feel coherent enough to draft a meaningful speech now. "I'm not sure what to say," I whispered.

She squeezed my shoulder. "I'll help you." Then she took to paper and pen and asked, "What do you want people to know about Cesare?"

I thought about it for a short while. "I don't want it to be sad, you know. I want it to be uplifting, kind of like a celebration of his life. I *want* to be uplifted by Cesare's life and not focus on the tragedy of his death, and I want others to do the same." My words seemed to run into one long-winded thought. "I think about how he was here for only forty short years. That's not very long, but he led a life full of love, success, and happiness. I mean really, when you think about it, he had it all. He had a great career that he absolutely loved, and he died doing what he loved. I mean, all those stories his colleagues shared about what a great pilot he was, how he mentored them, what a true professional he was. That honor guard—that was amazing. So much respect for him. They've been here for us throughout this whole ordeal. And they did that because of him, because they loved him." As I rambled on, I could feel the smile on my face. "And then there's this diverse group of ladies who loved him. He had a great family he loved, full of people who loved him back. He had a lot of grown-up toys, and he traveled the world. I can't think of anything he wanted to do that he didn't do. I've listened to all these incredible stories from all these people I don't know, strangers who shared with me these great anecdotes, and I'm comforted that he meant so much to so many. That he was so important in so many people's lives, you know. He really mattered. We should all be so lucky."

I looked over at the sun spilling in through the windows, and I felt its comforting warmth. "I'd like to focus on the lessons he left behind: to live life to the fullest just like he did and that it doesn't matter how long we're allowed to remain in this world; the important thing is how we choose to live our lives. That's the legacy he leaves behind." I stopped somewhat out of breath. "The problem is I don't know how to say all that."

Irene smiled. "You just did."

For the next hour, Irene took my rambling thoughts and helped me prepare a eulogy worthy of my brother. Then she handed me the piece of paper with scrawled-out words and edits in her handwriting.

The time had come to lay Cesare to rest. The same four funeral attendants who had accompanied us since Cesare had arrived at the funeral home solemnly walked over to the casket and started removing all the flowers. My heart beat wildly. Cesare's pallbearers surrounded the coffin and slowly rolled the silver box out the doors of the mortuary toward the cemetery. It had started to rain again. I was so sick of the rain.

We arrived at the burial site, where a white canopy covered a large hole in the ground. I read the words that bade my brother goodbye, my tears and the rain soaking the paper I held in my trembling hands. From the army of mourners came a hum of hushed sobs and quiet moans.

Then came the time to say goodbye, and my family members approached the casket together one last time. I kissed my yellow rose gently and placed it softly on Cesare's coffin.

"Good night, little brother. Sleep well."

PART II
The Whole Truth

"The way to right wrongs
is to turn the light of truth upon them."
—Ida B. Wells

7

Picking Up the Pieces

The thing about death is that life goes on. For the rest of us, that is.

The world continued to turn on its axis. Clocks continued to tick. The sun continued to rise and set. And the rain continued to fall. For seven days after the funeral, the rain fell on San Salvador in more sheets. Heavy, sloppy rain. Tears from heaven, everyone said.

For the next seven days, my family struggled to regain our footing, tried to figure things out, like feeling our way out of a room with no source of light. The house seemed dark and gloomy. I dragged myself out of bed every morning hoping a hot shower would help me feel something again. Family and friends delivered meals to our home, and we were deeply grateful for the kindness. I felt rudderless and needed to regain some sense of control, so I made a checklist. Freddie had to return to Los Angeles, but I stayed on for another three weeks to help with the myriad of things on my spreadsheet. Cesare's affairs needed order and that became my immediate purpose.

As fascinated as I'd always been with plane crashes, I'd never understood the complexity of dealing with the fallout of the trage-dies. I'd always assumed that placing one's loved one in the ground would be the end of the story. As an outsider, for me the story of a plane crash had always ended when I turned off the evening news. But I soon learned that as a family member of a plane crash victim, the story only began there.

After such a disaster, an investigation is triggered to determine the cause of the accident, so right away the crash site becomes inaccessible. That means everything inside the plane is held hostage until the investigators have collected the necessary data and compiled all pertinent information for their investigation. Marta Liliam informed us that Cesare's luggage, flight bag, uniform jacket and hat, cell phone, and personal belongings wouldn't be made available until further notice. This was very upsetting, as I imagined multiple sets of hands rummaging through his belongings. I knew Cesare would have been highly displeased with this. He was very particular about his things. I always thought he was borderline OCD, and I wouldn't have been surprised to learn he folded his clothes to specific dimensions. His luggage was always orderly; specific items went into specific pockets. Knowing strangers were haphazardly combing through his personal items and violating his privacy would have been torture for him.

In the days immediately following the crash, our family relied heavily on the airline to help us with all matters related to Cesare. They brought him home to El Salvador and handled all the necessary paperwork. TACA also delivered a thick manila envelope that contained the legal and medical documents needed to facilitate the transport of his remains. They oversaw the funeral arrangements and quickly processed Cesare's life insurance papers. I was grateful to TACA for easing our burden and allowing us to focus on grieving.

My priority at that time was to shield Mom from the media and the constant online speculation about what had caused the crash. It seemed everyone wanted to know what happened to the brakes, the runway, the plane. Every day, the newspapers remained splattered with the story and pictures of the broken plane. Evening newscasters continued to speculate, but there was nothing new to report, so the media kept the story alive by recycling the few known facts and filling in the gaps with conjecture. And suddenly, everyone was an expert. Eyewitnesses on land, passengers on the flight, people miles away who claimed to have heard the fatal boom. There were many theories. But the investigation had only just begun, so it was irresponsible for anyone to arrive at baseless conclusions. Many times I yelled at the TV, calling out the press for the unsubstantiated claims being spewed. At that point, nobody knew the whole truth.

At a press conference on May 30, 2008, hours after the crash, Manuel Zelaya, the President of Honduras, announced he was temporarily restricting airport access to allow only domestic flights and smaller aircraft to fly in and out of Toncontin while the accident investigation was underway. He never defined what he meant by "smaller aircraft," but he made clear that all commercial flights would be redirected to Palmerola Airport, a former US military airfield that had been constructed in the mid-1980s. "This is a decision that should have been made years ago but is urgent today because of the calamity that has occurred," he said. "Large aircraft attempting to land find themselves in these difficult situations not because of the location or length of the runway, but because the airport is in a valley surrounded by hillsides. The problem has never been the runway itself; the problem has always been the runway approach."

I was floored by his admission that Toncontin posed a problem for pilots. Because of its history of tragic crashes, for years people who lived around Toncontin had called for the relocation of the airport, but I was surprised Zelaya would publicly acknowledge the airport's safety shortfalls, especially at the start of the investigation. I also found it suspect that Zelaya dismissed the state of the runway as a possible cause of the crash. Did he know more than he was letting on? What information was he basing his conclusion on?

Zelaya was certainly right that incoming jets could expect a difficult approach when landing at Toncontin.

Inaugurated in 1948, Toncontin Airport was located four miles from downtown Tegucigalpa, the capital of Honduras, and was situated in a bowl-shaped valley surrounded by mountainous terrain. My research indicated that due to the adjacent mountain range, pilots approaching the airport were forced to make an unusually steep approach to clear the terrain. They had to then descend rather quickly while executing a dramatic forty-five-degree bank, then had to realign the plane by 180 degrees in order to land on its one runway. As I researched all day, I took copious notes and sketched data in my journal. Then every night during phone calls with Freddie, I took over the conversation, sharing everything I'd learned that day.

"Did you know pilots can land on this one runway in either direction? Runway 02 has a twenty-degree northeasterly heading perpendicular to the highway. That's the one Cesare landed on. Runway 20 has a 200-degree southwesterly orientation in the opposite direction. But essentially, it's the same strip of asphalt."

If I believed in God, I'd thank him for giving me my husband. There was no one else I could have had this conversation with. Freddie and I, both engineers, could geek out over infrastructure like no one else. "What about the runway length?" he asked. "Is there any info on that and how it compares with other airports?"

"The length of the runway is 6,112 feet, but that's misleading. You can't use all 6,100 feet because there's a displaced threshold."

"What's that?" he asked.

"Displaced thresholds are parts of the runway deemed unsuitable for landings. Their footage is included in the official length of the runway, but that number includes parts of the runway that can't even be used! They're designated for various reasons, like you might have obstacles immediately before the runway or the runway may have reduced structural strengths or there might be noise restrictions in certain communities. Toncontin has a 700-foot threshold! Planes have to land in the touchdown zone *beyond* the restricted threshold."

"So technically, this airport has an available landing distance of only 5,400 feet?"

"Yeah, 5,410 feet to be exact."

"Doesn't seem like much for jumbo jets. That's barely over one mile."

"Yeah. TACA 390 wasn't the first crash. In the last forty-six years, there have been nine aircraft accidents at Toncontin. That includes Flight 390. Sweetie, that averages to one aircraft accident *every five years*! All nine of the doomed planes were attempting to land, and five of those overshot the runway very similarly to Flight 390. In fact, on April 1, 1997, a US Air Force Lockheed C-130 cargo plane overshot runway 02. It rolled 200 yards before bursting into flames upon impact with an embankment and coming to rest on the major highway below. Three American military personnel on board died in that accident."

"Wow, that sounds just like 390. Eerily similar," he said.

"I know! And with all these crashes, with all these deaths, no one's ever done anything to address what screams of a dangerous airport," I said, shaking my head.

"It's so much easier to ignore the problem."

"How many more people have to die?"

Freddie didn't answer, and there was a long silence between us. Then he said softly, "Hey, listen, I hope you're getting some rest. I miss you."

"Miss you right back. Talk again tomorrow? I'll even let you tell me about *your* day," I said with a tired smile.

"Can't wait."

I hung up, then returned to my laptop. On these quiet nights while Mom slept, I kept myself busy poring over whatever information was available: press conferences, news releases, historical information. Burrowing myself in data and facts kept me from having to face my feelings. There was just no time for grieving at the moment. I would deal with my grief at a later time and on my own terms.

While immersed in my own inquiry, I learned the official accident investigation would be conducted by a committee comprised of various international representatives. This gave me confidence the investigation would be thorough and fair. The group included France's Bureau d'Enquetes et d'Analyses (BEA), the French counterpart of the NTSB; technical advisors from Airbus, the French manufacturer of the doomed Airbus 320-200; investigators from the United States' NTSB; an advisor from the Federal Aviation Authority (FAA); a technical advisor from International Aero Engine (IAE); and an investigator from Ireland because the plane held Irish registration.

Because the crash occurred outside the US, the Salvadoran Civil Aviation Authority assumed the lead. The committee's goal wasn't to determine fault or responsibility; it was to establish probable and contributory causes and to take steps to prevent a recurrence. It would be a long, drawn-out investigation.

At that point, the only source of official information available online was a three-page report prepared by the Salvadoran Civil Aviation Authority, titled *Preliminary Report – TACA International Airlines Flight 390, Airbus 320-233, Register No. EI-TAF, Toncontin International Airport (MHTG-TGU), Tegucigalpa, Honduras, May 30, 2008*. I downloaded the preliminary report, which provided the raw facts of the disaster but no conclusions, including the cause of the accident.

The report provided new details that filled in some of the gaps for the final twenty-six seconds of TACA 390. The black boxes had been recovered and taken to the NTSB for data extraction. The data regarding distances, speeds, and timeline added pieces to the puzzle and provided a greater level of clarity, like zooming in on a photograph. Touchdown occurred at 1,312 feet from the displaced threshold with a tailwind of ten knots. Flaps were fully extended, landing gear was down, and ground spoilers were armed. The nosewheel touched down seven seconds after the main landing gear and brakes were applied four seconds after. The aircraft overran the runway at approximately sixty-five miles per hour, dropped down a descending slope at the end of the runway, and crashed into an embankment.

I read the report several times looking for something beyond the narrative of facts and data. Perhaps I expected a mention of the "less than optimum braking conditions" Kriete had mentioned. Or maybe I wanted to see information about the status of the runway conditions at Toncontin. But I knew that at this stage in the investigation that was unrealistic.

The report was undated, which struck me as odd. This was an official technical report, yet there was no indication of when it had been prepared. In my line of work, the first thing we do when preparing a document is date it. It's more than just best practice; it establishes a timeline, a sequence of events. It provides precise and accurate information. It's a means to memorialize a historical event. So, when was this report prepared relative to the accident? How would we gauge the chronological order of events with other corresponding documentation? To my professional eye, the report looked unofficial, unfinished, amateurish. And it left me with a feeling of unease about where the investigation was headed. I suspected the final accident report would land on pilot error.

Since its construction and despite advances in aviation, Toncontin's runway had never undergone improvements. Yet, every year, planes got bigger, and the number of flights landing at Toncontin increased. Despite this and the airport's history of tragic crashes, the Honduran government never mitigated the dangerous runway conditions. The airport essentially remained the same.

This crash data history at Toncontin Airport should have

been alarming to the Honduran government, not only because of the high numbers of deadly crashes but because so many of the crashes had happened in the same way. During my career at Public Works, I spent a lot of time studying natural and human-made disasters. Our team planned, designed, constructed, and maintained public infrastructure of all kinds, including transportation facilities within our five local airports in Los Angeles County. We also collected and analyzed crash data that helped us plan roadway improvement projects designed to help eliminate traffic-related fatalities. In other words, we assumed responsibility for our infrastructure to ensure it was safe for the traveling public.

Zelaya knew there was a problem with Toncontin. He openly admitted it. He acknowledged the airport should have been relocated years earlier. So why did he wait for a human tragedy before taking action? Why did he wait for even more people to die before relocating to Palmerola Airport, a much safer airport? Why hadn't the Honduran government used Palmerola—with a runway length of 8,009 feet—as the airport of choice all along? Palmerola Airport's runway was about a half of a mile longer than Toncontin's. The distance between life and death.

The *Los Angeles Times* reported that for years, the Honduran government had been criticized for not abandoning the aging and undersized Toncontin Airport. Grieving families of plane crash victims, people who lived near the airport, and uneasy passengers all demanded to know why those in power hadn't relocated all international air services to an upgraded Palmerola Airport. But I already knew why. Money. Palmerola Airport was a military airbase. It didn't offer the perks and services of an international airport, like duty-free shops, restaurants, airline lounges, car rental services, and lots of parking spaces. So it didn't make financial sense to abandon Toncontin, an airport that was raking in profits. Palmerola would require a major investment to add adequate terminals and appropriate amenities worthy of an international airport, and it would be several years before the Honduran government saw a return on the investment. The bottom line was that it wasn't financially attractive to direct business away from Toncontin Airport.

For years, decision-makers had bet on a low probability of catastrophic failure. They decided the risk of crash was relatively low and didn't warrant action. To them, the loss of only a handful of lives every five years was a tolerable outcome. It was a deadly game of Russian roulette, and they were willing to keep playing.

Five weeks after the TACA 390 crash, Zelaya, pressured by the business community, announced the reopening of Toncontin Airport. As with all tragedies, people soon forget.

Despite our reliance on TACA to handle the high-priority issues surrounding Cesare's death, there were countless tasks our family had to handle ourselves. I was quickly overwhelmed by all the things that must be done when someone dies unexpectedly. We had to cancel his credit cards, freeze his bank accounts, make sure his life insurance policies were processed, and pay his utilities. And what about Cesare's cars? Had he driven to the airport that morning, or were both his vehicles sitting in his condominium garage? Did he have any outstanding loans? In our home, Freddie is the one who handles all our financial affairs—bank accounts, stocks, insurance. If something were to happen to me, I'm sure he'd know what to do. But what if something happened to Freddie? Would I even know where to start?

I'm the "doer" in the family, and I run my life with goals and milestones. Making lists and designating goals is also how I handle crises, by establishing order out of chaos. So I started with a checklist.

The first item on that checklist encountered a roadblock. In El Salvador, financial matters have to be handled in person. There was no 1-800 number to call to cancel credit cards or freeze bank accounts or cancel cell phones. So as our first order of business, Mom and I drove to Cesare's bank.

The security guard armed with the menacing machine gun opened the glass doors, and we walked into the air-conditioned bank, its ceramic tile floors shined and clean. We were greeted at the reception desk by a tall lanky young man who looked so young, I thought he must be an intern. Mom asked for the bank manager, and we were escorted past the line of customers waiting for the handful of tellers behind the glass windows, to an office toward the back of the room.

Mom spoke, "Hello, we're the family of Captain Cesare D'Antonio, the pilot who just died in the TACA plane crash in Honduras."

The bank manager on the other side of the desk winced. She was a young woman with a kind smile. "I'm so sorry for your loss," she said as she rose from her desk and came around to give my mom a hug. I was surprised by this gesture. It was something I wouldn't expect from my local Bank of America branch manager in Los Angeles.

I said, "We'd like to freeze Cesare's bank accounts and cancel his credit cards, please."

"Do you have the death certificate?" Her question was almost a whisper.

"No, we haven't received it from Honduras yet. We're only trying to freeze the account; we're not trying to access it or use his credit cards."

"I'm so sorry," she responded softly. "Only the account holder can freeze their account and cancel their credit cards."

"As I mentioned, Cesare is dead," I snapped while my mother sat crumpled in her chair, tears streaming down her cheeks. Out of the corner of my eye, I saw the tall young receptionist heading toward us with a glass of water. He set it in front of my mother.

The young woman across the desk shifted in her chair. She

said, "I understand, but it's bank policy. Only the account holder can freeze an account or cancel their credit cards. The only way a family member can freeze an account is by first producing a death certificate."

"Cesare's wallet and personal documents haven't been recovered yet." I leaned over the desk and lowered my voice. "We have no idea where they are or if anyone has them. We're just trying to prevent someone from using them." I continued to plead, but my patience was running low. "In the interim, we have all the paperwork from the transport of his body." I pulled out the manila envelope that held all the legal and medical documents we'd collected so far. "There are medical reports that declare him deceased. Perhaps you could use those for now while we wait for the official death certificate."

"I'm so sorry, but this is bank policy."

"Let's just go home," Mom said.

Mom had had enough, so I held her hand as we walked toward the exit with an air of defeat. The doors opened wide as we approached.

"Have a nice day," said the armed security guard, who looked just a little less menacing on our way out.

In order to properly deal with Cesare's financial and other logistical issues, it became clear we needed a death certificate, and that meant we were at the mercy of the Honduran government. For proof of my brother's death, we had to wait. And wait. It didn't matter that we'd placed him in a box and lowered him into the ground. Nor did it matter that the medical profession had declared him dead. Cesare wasn't legally dead until we received a death certificate that said so. I felt helpless.

On a rainy afternoon a few days after downloading the preliminary report, I sat at the dining room table to map out what I knew so far. Although my laptop sat next to me, I preferred to use pencil, paper, and ruler. I found it a satisfying way of organizing information as I tapped into my drafting skills acquired in my early college days. I was recreating Cesare's last seconds alive, and I felt the need to take my time, as if by performing this obsessive exercise, I'd be giving my brother back some of the time he'd been robbed of on his final day.

I drew runway 02 to scale, all 5,410 feet of useful landing distance, in a twenty-degree northeasterly direction. The preliminary report referred to the data using the metric system, the International System of Units (SI), most commonly used throughout the world. Although trained in both the SI system and US Customary Units, I converted all figures to the American standards units. There was something about speaking my own language I found comforting.

The preliminary report noted touchdown had occurred at approximately 1,312 feet from the displaced threshold, at which time the crew deployed engine reversers and ground spoilers, both intended to help stop the aircraft. The nosewheel touched down seven seconds after the main landing gear, and the crew applied manual braking four seconds after manual landing gear touchdown. Maximum pedal braking occurred ten seconds after that. Cesare then selected idle reverse to extend ground spoilers. At this point, there was only 625 feet of runway remaining, which the Airbus 320 covered in fractions of a second. The plane overshot the runway at sixty-five miles per hour, then collided with the sixty-five-foot-high embankment.

I placed little tick marks for each of these milestones on my runway drawing. I knew that with such a short runway, once engine reversers were deployed, TACA Flight 390 was committed to the landing. There was just not enough runway left for a plane to regain enough power to get back in the air, so the only thing left to do was to try to stop the plane. I looked at the little tick marks on my rudimentary sketch and wondered which one represented when Cesare knew he was in trouble.

As is the case all around the world, El Salvador's primary sources of information are radio, newspapers, TV, and social media. The media in all its forms controls narratives in order to sell more subscriptions and secure funding from sponsors, and long gone are the days of trusted news delivery because misrepresenting the truth has become lucrative and common practice. Perhaps one of the most dangerous byproducts of such disinformation is that people parrot the information disseminated by the media regardless of whether they know it to be true.

The day after Mom's and my futile trip to the bank, Beatriz, a distant relative, stopped by to visit Mom. I took the opportunity to escape to my room for some time alone. I laid my head on a pillow and could overhear their conversation, muffled through the concrete walls. The ceiling fan made a whirring sound that I found soothing, but as soon as I began to relax, their voices began to boom. Shortly afterward, I heard the front door close, and Mom rushed into my room with tears streaming down her face.

I sat up straight. "What's wrong?"

"Beatriz is saying they determined it was pilot error."

"*What*? Who the hell is 'they'?"

"The newspapers. The newspapers are saying it was Cesare's fault. Is that true? Rossana, is that true? Tell me!"

I hugged Mom tightly, as if trying to muffle the words that still rang in her head. Mars and I had tried desperately to protect Mom, and part of that protection meant we'd agreed to not discuss the accident with her. It was too soon. It was too raw. It seemed we'd all silently agreed that these early days were for mourning, supporting each other, talking about Cesare, and helping each other through the pain. I'd jotted all my furious research quietly in my journal, sharing my findings only with Freddie. So to now have someone, anyone, violate that unspoken mourning-time understanding enraged me.

And not only did someone have the gall to come into our home and compound Mom's pain by gossip mongering, I, only a few feet from my mother, had failed to protect her.

I pulled her away so she could see my face, hot with rage. "Look at me. What the hell does Beatriz know? Was she on that plane with Cesare that day? No. She's an idiot, an ignorant idiot! She doesn't know what the hell happened that day. Not even the experts know."

"Then . . . she shouldn't say these . . . things to me," Mom said, shaking her head.

"Exactly. Don't listen to what anyone has to say, Mami. Listen to your heart. Cesare saved the lives of those people on that plane. That's all that you need to know. Anybody who says different doesn't deserve to be in this house." I held her convulsing body again as she buried her head into my shoulder. Then I said slowly and quietly, "If anyone ever says something like this to you again, you need to ask them to leave. No questions. No explanations. They are not welcome here. You tell them to leave. Period. Promise me."

"I promise," she pledged, her voice shaking with the fragile timbre of a lost little girl.

My heart broke all over again. Mom was exhausted. She wasn't eating or sleeping. In just a few days she'd aged years. Her face was now lined with deep creases that flooded easily with streams of tears, and her tired skin sagged. Gone was the vibrant woman with the wicked smile. Mom lay down on the bed and almost immediately drifted off, broken by exhaustion and grief.

I stomped to the living room and grabbed the phone, flipping through the address book until I found my cousin's number. Then I paced back and forth on the patio where I was confident my mother couldn't hear me. As I listened to the phone's ringing, my heartbeat thundered.

"I'd like to speak to Beatriz, please. Yes, this is Rossana." I surprised myself with my politeness. Interesting. On the outside I was behaving with great self-control, while inside, I was screaming, "Patch me through to that insensitive bitch, pronto!"

"Hi, Rossana, how are you?" Beatriz's voice was perky and light. She had no idea the storm that was headed her way.

"Not so good. Listen, I'm not really sure what happened when you visited earlier today, but whatever you just said to my mother really upset her."

"What do you mean?"

"I don't know if you know this, but there's an accident investigation that's just getting underway. So whatever happened on the day of the crash is still unknown. *No one* knows what happened, not the media nor the experts, much less you. I don't appreciate you coming into *my* home and feeding my mother misinformation about what *you* believe to be true. And just to be clear, regardless of what the investigation concludes, Cesare will always be a hero to Mom and my family. Do you understand?" I felt the need to enunciate every word as if dealing with a child who was struggling to keep up with me.

There was a short silence before she responded. "Rossana . . . I'm . . . sorry . . . I never meant to upset your mom, really . . . I was just repeating what we heard in the news."

"It doesn't matter what the media says. You do not, *under any circumstances*, make irresponsible comments like that to a grieving mother, for God's sake! You were way out of line! I'm going to ask you to please not call here again. Please don't visit, either. If you want to attend the mass, you can do so, but please do not approach my mother. You've done a lot of harm, and it's going to take some time to fix it."

Beatriz replied in a cracking voice, "I'm so sorry. I will not bother you again."

"Good. Goodbye." I slammed my thumb on the TALK button, ending the call, ticked off that technology had robbed me of the satisfaction of slamming the phone down.

I'd been pretty rough on Beatriz. I knew I'd pushed her to tears, but I didn't care. At this point, I didn't care who else was hurting.

Unlike with other crashes that occur in remote locations, the black box for Flight 390 was readily accessible and easily retrieved within the first three days of the accident. The term *black box* is a holdover from the days of World War II. In fact, in the 1960s, the International Civil Aviation Organization (ICAO) required that flight data recorder and cockpit voice recorder boxes be designed using fluorescent orange, so they'd be easier to find following accidents. Per the ICAO, recorder boxes are designed to withstand high-speed impact and intense fire conditions and are typically located in the rear of the fuselage to increase the likelihood they'll remain intact after a disaster.

In the first two weeks after the crash, I stumbled across a link to the cockpit voice recordings posted online. The link included an audio file and the transcripts. I hovered the mouse over the audio hyperlink, the palm of my hand moist, my pulse racing. What might I hear? These would be the final moments of Cesare alive. His final words. The sounds of death itself. I couldn't bring myself to listen. But I wanted to know what had been said, so I downloaded the transcripts of the conversation between the cockpit and control tower, then closed the website.

"What did you learn today?" Freddie asked, several hours later. He sounded tired. I hadn't asked him how he was holding up or how he was doing at work. I hadn't asked whether he was eating or sleeping. For several days, I hadn't asked about any of it. But there would be plenty of time for me to dwell on that branch of the guilt I was feeling. I'd add *neglecting my husband's needs* to the list of personal offenses I was racking up, along with *not protecting my mother* and *bullying my cousin*.

"I got ahold of the cockpit voice recorder transcripts. They spell out the communications during the final thirty minutes of flight."

"Oh, wow. Are you okay? You wanna tell me about it?"

"Well, there's some technical jargon I don't understand. I'm gonna have to research it. But you know, the transcripts verify a lot of what we already knew. And as you can imagine, the activity in the cockpit is very intense toward the end of the recording, including overlapping words that lead to unintelligible information, which could have led to miscommunication," I said.

Freddie was quiet on the other end, letting me process. He knew this was hard for me. Then he said, "Oh, babe, I wish I could be there to hold you." Tears began to well in my tired eyes, but I snapped myself out of it.

"There's something else," I said. "The timeline seems a little off. I mean, every time someone says something or a sound gets picked up by the mic, there's a line of text with the corresponding time stamp. But it's showing there are often several minutes between each text entry. I'm not sure I'm making any sense."

"You're saying it's not a realistic timeline. It only shows activity. But it leaves out silences. Silences that could be several seconds or minutes long."

"Exactly!"

"So if you were to expand the timeline to include those silences, you'd have several lines of no text, which would offer a more accurate representation of what was happening at a microsecond level."

"Right!" I wanted to marry him all over again.

"But what would an expanded timeline tell you?"

"It would give me a more-detailed chronology that would show the gaps in conversation. I'd be able to literally *see* the silences, the ebb and flow of activity, and any overlapping speech."

"That's interesting. You're right, pilots aren't in there just twiddling their thumbs. There's a lot going on in that cockpit—real-time weather conditions, a change in runway . . ."

"There's something else that struck me as odd. Before Cesare tried to land on runway 02, he radioed TACA operations and asked whether Flight 390 should continue to the alternate airport in case they couldn't land at Toncontin. Doesn't that strike you as odd? I mean, you'd think there would be some written procedure, some policy that established the protocol whether to proceed to an alternate airport if the pilot was unable to land. Why would he ask for instructions? Cesare had already landed at Toncontin fifty-two times. But on that day, he asked for guidance. The only reasonable explanation is TACA had no clear policy for this type of event."

"Or a policy that wasn't necessarily enforced consistently. Listen, I have an early meeting tomorrow. How about you take a break and get some rest?"

"Okay, love ya."

"I love you more than that."

Despite the thousands of miles that separated us, Freddie had managed to lighten my load. I smiled as we ended the call and thought about how lucky I was to have found that husband of mine.

I couldn't stop thinking about Cesare's question to TACA operations. Did his query expose a lack of policy? Or was there a preestablished procedure that wasn't consistently enforced? I decided the issue was probably moot at that point, except to shed light on the true culture at TACA. A culture of safety is something created and enforced at the highest levels of a corporation's leadership, and there should be no question for frontline workers about where the organization stands on the implementation of safety procedures. TACA Airlines was well aware of the shortfalls of Toncontin International Airport, yet they accepted this level of risk and an environment that posed safety concerns with each takeoff and each landing.

I sat in the quiet of my mother's home looking at my sketch with the timeline tick marks, and I could just feel it in my stomach, that despite the alarming facts about flaws in the runway and the airline's high-risk tolerance, Cesare was going to be made the scapegoat.

8

Lost and Found

It was the end of June 2008, four weeks after the crash, and I was scheduled to return to Los Angeles the next day. Marta Liliam called to let us know TACA had recovered most of Cesare's personal belongings including his carry-on, his uniform jacket, and his wallet with all his credit cards and $80 in cash.

TACA arranged for Cesare's belongings to be delivered to his condominium, and I took a taxi there from my mom's house to collect them. As I rode up the elevator to the fourth floor of Cesare's building, I could feel my pulse rise in my neck. Except for Galo who had picked up Cesare's uniform on the day of the crash, no one else had been here. Cesare had bought the place about five years before, his first real estate investment. After signing on the dotted line, he'd called me, excited to share the news. "Hey, I'd love to show you the place," he'd said. "When will you be in town? Don't forget, I'm expecting a generous housewarming gift!"

The newly constructed condo was located in a modern building in the foothills overlooking San Salvador. It was in a new part of town that had seen a boom in high-rise residential buildings. Cesare was now a proud homeowner; I knew he'd spent much time and money on making it his dream home.

What struck me when I first walked into the living room was the scent, a fragrant blend of cedar, amber, and musk. It was as if Cesare was in the room. I felt a shiver and my fingertips tingled. The layout and design of the living room was something straight out of *Architectural Digest*. The only other time I'd seen the place, Cesare hadn't yet fully furnished it, but on this day, it was decorated and inviting. Cesare had always been a very particular bachelor. In fact, when it came to things like the need for symmetry and orderliness, I thought he might even be a bit obsessive-compulsive. The tile floors sparkled, and the cushions on the sofa were puffed up and lined up with their corresponding V-shaped dents pointing up. I could just imagine Cesare walking around the living room karate-chopping each of the decorative pillows to give them maximum fluff. The whole place looked as if it had been readied for someone to come home after a long day. But Cesare wouldn't be coming home again.

I ran my hands along the chocolate-brown leather sofa, then walked toward the cherry wood entertainment center against the wall. Family and friends smiled back at me from picture frames neatly arranged on the shelves. Happier moments frozen in time. Maroon-colored vases in diminishing sizes stood guard on the coffee table accented by never used scented candles. Vanilla bean. His favorite.

I opened the sliding glass doors and admired the view of the city, a majestic volcano on the horizon. The volcano's name escaped me. From up high, there was a quiet serenity, and I wondered when Cesare had last enjoyed this amazing view.

I stepped back into the living room and caught sight of his bar, fully stocked with bottles of all shapes and colors, the labels all facing the same direction. The gourmet kitchen sparkled. He loved to cook and had become an impressive amateur chef. But just like a brother, when I'd last visited, he'd ordered takeout. Pizza. He'd sauntered out of the kitchen holding the large box.

"Pizza?" I whined. "A gourmet kitchen and I get pizza?"

He grinned. "Next time I'll cook you a three-course meal!"

Next time never came.

I entered Cesare's bedroom. The bed was so neatly made,

I knew I could have bounced a quarter off it. The only thing that seemed out of place were his sneakers on the floor near his bedroom window, probably left there after his morning run on the treadmill. His clothes hung neatly organized in the walk-in closet—shirts followed by jeans and shorts, then dress slacks, then suits. At the far end were his white, long-sleeved work shirts with the tabs on the shoulders for epaulets, the four gold bars that distinguished him as the captain. When Cesare had worked a part-time job as a flight instructor while waiting to be accepted at the airline, he could afford only one white work shirt, and Mom washed and ironed that one shirt every night. Now, fifteen years later, there were at least twenty crisp white shirts hanging in dry-cleaning bags, ready to report for duty.

On a shelf in the corner of Cesare's closet was his stash of cigarettes—Marlboro Reds. The ones with the cowboy. I picked up a pack and took a strong whiff. Cesare smoked until the day he died. I'd quit twelve years before, but on high-stress days I missed taking a long drag. As children we'd seen our dad wither from lung cancer, his life extinguished, so it made no sense that we wouldn't have been scared away from smoking, but we both picked up the nasty habit. When I decided to finally abandon the poison sticks, resorting to nicotine patches to offset my withdrawal symptoms, Cesare had been my staunchest supporter.

As I stood in the middle of my brother's room, I felt like an intruder, a voyeur in this private space that would never again be occupied by its rightful owner. I wondered what he'd want me to do with his belongings. What about this condominium? His beloved Porsche. Clothes and wall hangings. Home accessories and trinkets. Gym equipment and kitchen appliances. All of it. Where should all of it go? I had no idea, but who else should the burden of figuring it out fall on? Who else but me should tie up his loose ends? *Take care of your little brother. Always take care of your little brother.*

I headed back to the living room, sat on the floor, and slowly unzipped the travel bag delivered by TACA. I took my time, grateful to be alone in his space, although I somehow sensed my brother's presence throughout. There, neatly folded on top of all the other contents, was Cesare's jacket, the core of his uniform. It was the one he'd been wearing that morning as he crossed the threshold into the jet, black with the four gold stripes on the sleeve cuffs. Once inside the cockpit, Cesare would have taken it off and hung it in the wardrobe closet adjacent to first class. We were told the jacket had been dry-cleaned, and my mind raced as I imagined what kind of soiling had occurred that called for cleaning. But I shook the dark thoughts from my mind.

Under his uniform jacket, I found a small jewelry box. Inside it were his flight wings and a little lapel pin, a gold and crystal angel. Marta Liliam had told our family Cesare had been wearing the little crystal angel pin just below the flight wings on his jacket that day. I was brought up believing angels were symbols of guardianship and protection. It was all part of the Catholic machine Cesare and I had grown up at the mercy of. All those symbols and songs and chants and prayers and rules that were supposed to protect. I stared at the little crystal angel that lay in the velvet lined box and wondered where its magic powers of protection had been four weeks ago.

What would the priests say about that? I thought. But I didn't need to ask. I knew they'd hand me some bullshit about how "God has a plan." "Everything happens for a reason." "The mystery of his will." It's the perfect system: We're taught that when something wonderful happens, we should thank God for it. When we experience something unthinkable, horrible, or unjust, God is testing us, and we're supposed to just accept it. The original abusive relationship.

I reached into a large plastic bag and found what was left of Cesare's Palm Pilot, its screen broken into little pieces, its frame hollowed out. In another plastic bag were the deeply scratched frames of his Prada sunglasses. Only the frames, no lenses. He most likely hadn't been wearing them due to the heavy fog that morning. So how did they break? Did they fly out of his shirt pocket?

Inside a green plastic bag were the black, slip-on loafers he'd been wearing. They were torn, and the stitching that held the saddle across the top of the shoe had come undone. What could have possibly torn them? I frantically looked for signs of blood but couldn't find any. I pressed the shoes against my thundering chest.

His iPod sat intact at the bottom of his carry-on. Without thinking I tapped it on, and the screen sprang to life displaying the white Apple logo with the bite missing. Moments later, David Cook's scruffy face stared back at me. Cook was a rock singer and an *American Idol* winner, and the song onscreen was *The Time of My Life*. The iPod would forever play this anthem that seemed to portray Cesare's life in its haunting lyrics. Cesare had been living the time of his life.

Cesare had packed enough clothes for two days in Miami: shorts, T-shirts, and flip-flops. Perfect attire for the rest and recreation he must have been looking forward to and never got to enjoy. To touch and hold items my brother had so recently touched made me feel closer to him. Unable to hold Cesare, I grasped what remained of him.

In tragic events like these, families find comfort in the smallest of things that under normal circumstances are taken for granted. We were fortunate to have most of Cesare's belongings returned to us, although we never recovered his cell phone or his beloved jewelry—a Cartier watch, a gold necklace, and a ring he wore to conceal a Celtic tattoo on his ring finger. I'd never asked him about the significance of the tattoo, and now I wondered why.

In those first few days, these missing items caused agonizing frustration and stress. "Who has them?" Mom asked frantically, again and again. She even considered designating someone to fly to Honduras to search for them. But in just a few weeks, we accepted the items were probably lost forever. They might have been taken by someone involved in the rescue, and I became enraged every time I thought about someone stripping precious personal items from the body of a downed pilot but not returning them to the people who loved him.

I've thought a lot about other disasters. Disasters after which families were never able to recover their loved ones or the people's belongings. Missing children who disappeared without a trace, 9/11 victims whose bodies were pulverized, soldiers whose remains never returned home. I think of these families often. And in that context, I remind myself we're the lucky ones because we had Cesare returned to us. His body broken and lifeless, but Cesare, nonetheless.

In my peripheral vision I saw a flicker of movement and looked up to see a spotted butterfly land on the enormous TV near where I sat. It was a beautiful creature with regal wings of orange and yellow, outlined in black. I sat silently watching its soft wings flutter. *Where did you come from?* I wondered, then glanced out at the fourth-floor balcony, its sliding door wide open, a sheer white drape billowing in the breeze. Spellbound by the little creature's beauty, I was reminded of the butterfly effect, the notion that the fluttering of a butterfly's wings in one place can trigger currents that might ultimately grow into a storm even half a world away and dramatically affect events nowhere near where the butterfly sits. In other words, everyone and everything are connected.

I sat back against Cesare's smooth couch and let my mind wander about this idea of cause and effect that had always fascinated me. Many of our choices have no significant consequences as far as we know. Chocolate or vanilla. Stripes or plaid. A shortcut home or the scenic road. But the butterfly teaches us that every decision creates a ripple. Cause and effect. Every day, each of us is a butterfly flapping our wings, and in so doing, not only drastically affecting our own lives but maybe even changing the world. A six-year-old girl who, prompted by her classmate to run, heeded the warning and managed to escape the Sandy Hook school building while the deranged killer reloaded his AR-15. As she fled the scene, the madman sprayed another blast of bullets, killing twenty innocent first graders, including her brave little friend who screamed, "Run!"

The 9/11 victim who wasn't going to fly to Los Angeles that day but who at the last minute changed her travel plans so she could be with her husband to celebrate his birthday. This change of heart secured her a seat on American Airlines Flight 77, and shortly after takeoff, she was killed as her plane was crashed into the Pentagon.

A thirteen-year-old girl walking home along the Pacific Coast Highway after a sleepover in Malibu arrived at an intersection at the exact moment a crazed, out-of-control driver fleeing authorities swerved off the road. He plowed into the child, which launched her body into the air, killing her instantly.

Randomness. Change one thing, change it all.

As I sat in my brother's living room, I imagined what his last day had been like:

The alarm went off at about 5 a.m. He didn't hit snooze but jumped out of bed groggy and headed for the closet to dress in shorts, T-shirt, and sneakers for a morning jog on his new treadmill. Half an hour later, sweaty and alert, he threw his workout clothes in the hamper and set the sneakers near the bedroom window. After his morning shower, he trimmed his beard and brushed his teeth. As usual, he skipped breakfast. Applying a generous dose of cologne, he left the bathroom and headed for the walk-in closet, then stripped the dry-cleaning bag from his captain's uniform and dressed meticulously, snapping the captain's epaulets on the shirt's shoulder straps. At about 6 a.m., he grabbed the case that held his flying manuals, suitcase, and cap and proceeded down the elevator to street level where he boarded the airline shuttle for the forty-five-minute ride to the airport. Rain danced heavily on the shuttle's windshield. On his way to the airport, Cesare phoned Mom and told her he was flying to Miami. It was a short conversation. As always, Mom gave him her blessing and told him she loved him. At approximately 7:30 a.m., he boarded the plane and took off in heavy rain in what newscasters reported as Tropical Storm Alma. A few hours later, he was dead.

My heart beat wildly as I pulled myself out of the daydream and back to the reality of Cesare's living room. I stared at the beautiful butterfly, still at rest on the TV, and wondered which of its distant cousins with fluttering wings, perhaps clear across the world, had set Tropical Storm Alma in motion. If any one of the events here or anywhere around the world that morning had gone differently, would Cesare still be alive? What if Cesare's alarm clock had failed to go off? What if he'd strained his back on the treadmill? What if he'd selected the other uniform that hung in his closet? What if he'd driven himself to work instead of taking the shuttle? What if? What if? What if?

The curtain billowed once more, and the little butterfly flapped its wings, then flew past the sliding doors, outside, and up into the cloudy skies. I watched him fly away until he was out of sight and wondered what wicked storm that beautiful little creature had just set in motion.

The next day, after four weeks in El Salvador, I was on a plane headed back home to Los Angeles. Sitting in seat 3A of first class, I was grateful the seat beside me was empty. I wanted to be left alone.

I stared out the window at the puffy white clouds that blanketed the sky. The sun was intense, and it warmed the tears that streamed down my face.

Years earlier, just after I'd been assigned my very first office with a window, I'd been out to dinner with Cesare and boasted about my new office as the ultimate sign of corporate success. He'd rolled his eyes and asked, "What's the view?"

"What do you mean?" I'd asked.

"What do you see when you look out the window?"

"The parking lot," I admitted sheepishly.

He smirked at me. "When you look out the cockpit window—now *that's* the ultimate corner office view! What other profession lets you fly through the heavens?"

Today, I knew exactly what he meant. I wished I could tell him I now understood how right he was.

The cabin steward announced the movie was about to start. The on-flight entertainment was the movie adaptation of Gabriel Garcia Marquez's *Love in the Time of Cholera*. Only six months earlier, Cesare had given me the book for Christmas.

Many holiday seasons before, I'd suggested all my family members prepare wish lists to make the other family members' lives easier at Christmastime. One year, after carefully studying the Tiffany catalog, I included on my list several little overpriced Tiffany trinkets. Cesare was the first to balk, "What? Do you think we're all made out of money?"

Underneath our family tree that year, amidst all the other beautifully wrapped gifts, was an unmistakable Tiffany blue box tied with a beautiful white ribbon. The card read *From your brother, who you also don't deserve*! I opened the box to reveal my much-anticipated silver bracelet.

Now I ran my fingers along the silver links on my right wrist as I stared blankly at the small movie screen before me. For our last Christmas together, my list had been more modest. Cesare had given me the Garcia Marquez classic and a Burberry Brit perfume. The perfume had run out just a few days ago, and I couldn't bring myself to discard the empty little plaid bottle. I had tucked it in my carry-on, and it was on its way home with me. But where was the book? I'd read it shortly after the holidays, then carelessly discarded it somewhere in our office at home. Where had I put it? Did I give it away? Did I loan it? Did I lose it? Where was it? I made a mental note to look for it as soon as I arrived home.

We hit an air pocket. My body tensed, and I clutched the handrests on either side of me. *Breathe in, breathe out. Breathe in, breathe out. Breathe in, breathe out.*

Focusing on the peaceful clouds outside my window, I soon felt at ease. "Cumulus," I whispered to no one and felt my muscles relax as I was transported back in time.

I was ten years old. Cesare was six, and the two of us were sprawled lazily on freshly cut grass watching the clouds drift across the blue sky.

"That one looks like a hippopotamus with a birthday hat!" he said.

"Where?" I asked. "That's not a hippo! That's a fire engine with its ladder up!"

He pointed to another part of the sky. "Look! Over there! A chubby baby in diapers!"

"Yeah, it looks like you."

"Does not!"

"Does so."

"Mom!"

As if on fast-forward, the scene advanced several years. Cesare was studying to be a pilot and was working on his theory classes. It was a bright sunny day in the San Fernando Valley, and the intense blue sky was dotted with billowy clouds.

Cesare pointed to the skies. "Do you see those puffy cotton-like clouds over the treetops? Those are called cumulus clouds."

"Okay," I said, distracted.

"They're usually found at low altitudes and produce very little rain."

"Uh-huh."

"Cumulus congestus clouds are different. They look like a narrow tower, and they can grow into cumulonimbus clouds. *Those* can produce rain."

I was only half listening, thinking a cloud is a cloud is a cloud.

"Are you even listening? What did I just say?"

"What? Of course I'm listening. You just said it was gonna rain."

Clearly, I'd failed my lesson on clouds. Or had I? Reflecting on that day, I realized his words had sunk in and tucked themselves away somewhere in the recesses of my mind. The information had seemed useless at the time, yet today, it was a gem, a piece of Cesare left behind. I looked out the window at the cumulonimbus clouds that seemed to be escorting our metallic bird home. I was still smiling when the captain announced the start of our descent.

9

Fly Like an Eagle

I 'd been back in Los Angeles for a couple of months, and it was good to be home. I'd hoped placing some distance between San Salvador and me would provide some peace, but I soon realized my grief wasn't tied to a location. I carried it with me, and it was all consuming.

I'd returned to work at the "cube," as we referred to Public Works headquarters in Alhambra, a thirteen-story glass building. On my first day back, the automatic glass doors had swung open as if greeting me with open arms, but I really dreaded being back. I knew curious minds would poke and prod for information about my brother's crash, and I'd have to relive the ordeal all over again. I had worked at Public Works for eighteen years, almost half of my life, but it now seemed like a foreign land. Trying to avoid a crowd of people at the elevators, I headed for the stairs and trudged up to the third floor, each step painfully heavy on my already shaky legs, pausing at each landing to catch my breath. Finally, I reached my floor, took a deep breath, and stepped onto the carpeted cubicle farm, catching sight of heads peering over the workstation walls as I walked by. As I navigated the maze of workspaces to my office, I overheard hushed whispers and could sense sad, piercing eyes follow me as I walked. What was there to see? The stares felt like the morbid fascination with car crashes that compels us to look despite what we might see. I myself had fallen prey to this curiosity. This time I was the spectacle.

My boss greeted me at my office door. Dennis had never been the sensitive type, but he reached out and wrapped his arms around me, my briefcase hanging limply as I tried awkwardly to reciprocate.

"Welcome back," he said softly, his eyes welling up. A handful of colleagues stood behind him. As much as I wanted to hide in my office, I knew this welcome was far from over. But I quickly realized that, as bad as I may have felt at that moment, this family of sorts was struggling with what to say. They were hurting for *me*. So I did what I do best. I tried to make each of them comfortable with the uncomfortable situation.

"Thank you, I'm fine."

"Yes, it's good to be back."

"Absolutely. Work will be good for the soul."

But work didn't offer the escape I'd hoped for. I took every opportunity to hide in my office. I attended meetings faithfully and hoped I'd make rational decisions but often caught myself staring blankly at the computer screen as the minutes and days ticked away. I looked forward to the upcoming Labor Day holiday, a three-day weekend at home, just Freddie and me.

That Saturday afternoon, I suggested to Freddie we drive up the coast. Always a trooper and perhaps happy about my willingness to get out, he said, "Of course!" An hour later we were on our way. The ride offered beautiful Pacific Ocean views, white-capped waves stoked by a strong ocean breeze. The winding road we chose cradled us between stunning bluffs and a wide expanse of crystalline waters. We rode in quiet, and I gulped the salty and tangy ocean air.

When it came time for lunch, we stopped at the quaint little outdoor café at Camarillo Airport for sandwiches and lemonade and sat outside enjoying the clear blue skies and soft breeze as we watched little planes take off and land on the airport's only runway.

The Camarillo Airport in Ventura County, just west of Los Angeles, is considered a reliever airport serving privately operated and executive aircraft, which means there's no commercial service. Curiosity got the best of me, and before leaving the house I'd switched on my computer and checked the airport's stats. The little reliever airport's one runway was 6,010 feet long, six hundred feet longer than Toncontin's.

Several planes were parked on the tarmac, and after lunch Freddie hung back while I walked out to the airfield for an up-close look at the little flying machines. Cesare had learned to fly in one of these two-seater Cessna planes, and I wanted a moment alone to reminisce.

Years earlier, Cesare and I had been roommates while he studied to be a pilot. In order to obtain his commercial pilot's certification, he was required to complete a cross-country flight of no fewer than three hundred nautical miles and land at three airports with operating control towers. He explained that the term *cross-country flight* was misleading and there would be no crossing the country but that the FAA defines a cross-country flight as one with a minimum of fifty nautical miles between departure and destination airports. He'd already earned his private pilot license, so he suggested inviting a few friends and making this flight a day trip to the Grand Canyon.

Our friends accepted the offer with excitement, but as the day approached, they flaked out one by one, plagued by mysterious last-minute plans or appointments they couldn't get out of. My hunch was they were all too scared to fly in a little plane with a novice pilot, and I think Cesare believed this too. I resented them for it, for leading him to believe they trusted his skills only to back out at the last minute. But I probably would have done the same had it not been my brother at the controls.

A week before takeoff, Cesare and I sat at the dining room table of our apartment and finalized our plans. With the map splayed out before us, we decided to skip the Grand Canyon and instead planned a day trip to San Jose. We would depart from Van Nuys, land in San Jose, then at the end of the day stop in Burbank on our way back to Van Nuys. We'd cover more than three hundred miles and land at three airports. I would be his first passenger. My claim to glory.

The first thing I noticed when we walked onto the tarmac at Van Nuys Airport was that the planes looked awfully small. I assumed it was a matter of perspective and that the planes would look much more impressive as we got closer to them.

"Here she is," Cesare said proudly, pointing to a Cessna two-seater. It was a cream-colored plane with brown stripes along its sides and wings. It looked like a toy, an oversized remote-controlled plane I'd expect to see kids flying across a parking lot. Metal link tiebacks hooked the wings to the concrete tarmac. All the little planes seemed to have them. "What are the chains for?" I asked, even though I was pretty sure I already knew.

"To hold the plane in place," Cesare said matter-of-factly, his upper body leaning into the cockpit as he fiddled around with gears or controls or something else important.

I scanned the army of tiny planes chained to the ground to keep them from being blown away in the wind and felt a little shiver. "I don't know if this is such a good idea."

"Oh, come on," he said, "you're gonna love this."

I wondered if he was nervous. Was he at all nervous because I'd tagged along? Did a pilot feel any different when he flew with a passenger? Did it make him more careful, like so many mothers driving with *Baby on Board* signs in the back windows of their cars?

Fuel. Check.

Wind. Check.

Flight record. Check.

Instruments. Check.

"Hey, you forgot to kick that tire!" I joked. A nervous giggle gave me away. Cesare rolled his eyes.

No more stalling. I hopped onto the plane, buckled in, and held on as he drove the little Cessna to the airstrip and aligned it with the center of the runway. We finally got clearance for take-off, and the plane barreled down the runway, me bouncing in the passenger seat as we gained speed.

"Power available. Air speed alive." Cesare called out. He then pulled back on the yoke, and the plane left the ground and took flight. Once airborne, I had a sinking feeling, literally, as the plane's thrust pushed me back into my seat. It felt like being on a roller coaster. The plane vibrated as it strained to gain altitude and the ground began to fall away. Holy crap! My little brother was actually flying this contraption.

I peered out my side window and watched the buildings seem to get smaller and smaller, then felt a slight wave of nausea. We were still ascending, and the little plane rattled like a Mexican maraca. I'd heard pilots often describe a sense of peace and serenity while in flight, but I wondered how that was possible with all the clattering.

We finally stopped climbing and the plane leveled off. "Okay, we're on our way headed north flying at an altitude of about 9,000 feet," Cesare said, sounding like an actual pilot.

I said, "Where's the flight attendant? I need a drink."

"Relax and enjoy the ride."

The rattling had now subsided to a hum, and it was somewhat soothing. Something about the noise now being steady and even helped slow my pulse.

The ground below came alive in vibrant colors that ranged from ocean grays to crop-field yellows. Mountain ranges in bold relief popped up off the low-lying valleys. The silhouette of a cargo ship appeared to stand still on the horizon. It was all majestic and breathtaking.

Cesare banked the Cessna ever so slightly, and I caught sight of the deadly San Andreas Fault, its jagged fissures scarring the face of Mother Earth. Just like Cesare had lectured me about cumulus clouds years before, it was now my turn to rattle on about tectonic plates and fault zones. "Did you know that the southern segment of the San Andreas Fault hasn't ruptured in over three hundred years? We're overdue for a big one."

"Then it's a good thing we're flying north!"

A few hours later, Cesare announced to the control tower we were approaching San Jose International Airport. I was glad to hear we'd be landing soon and I'd be able to stretch my legs, finally released from this little plane that looked like a toy. The radio cackled, and the voice of an air traffic controller welcomed us to San Jose. I was surprised when Cesare, still looking out the window, said, "Okay, we should start looking for the airport."

"*Looking* for the airport? Are you talking to me? You don't know where it is?" I asked with exaggerated panic.

He ignored my remarks as he scanned our surroundings. I looked down at the map strapped to his leg and thought of a cartoon map in an amusement park with a big red dot that reads, *You are here*. I silently hoped Cesare's map had on it a big red arrow pointing to a cartoon airport.

I looked out the cockpit window, hoping my wide-open eyes would increase my sphere of vision, frantically scanning the blue skies, wondering what the hell I was looking for. How would I know if I saw it? Would there be a big neon sign somewhere with big letters spelling *AIRPORT*? Perhaps a Las Vegas–style blinking arrow directing us? Maybe a Jetsons-type hologram? *Airport, left at the next cumulus.*

I could see Cesare's forehead was pinched as he surveyed the skies through the windshield. *Maybe if we followed another plane,* I thought. But there was no other plane in sight. Where were all the planes, for God's sake? We were headed for an international airport, weren't we? San Jose International Airport, that's what the map strapped to Cesare's leg said. What kind of international airport was this? Where were all the other jets? Where was all the activity?

Finally, Cesare pointed slightly northeast. "There it is."

My eyes followed his index finger, but I couldn't see anything. There was no big neon sign. No arrow. No blinking lights. Nothing! But Cesare seemed to be in control. Trusting he knew what the hell he was doing, I held on as he steered the little plane in that direction.

The radio came to life once again and a voice guided Cesare and our Cessna to the landing strip. As we prepared to land, Cesare performed his pre-landing checklist. Brakes, check. Undercarriage down and locked, check. Flaps, check. Master on, check. Our little Cessna vibrated and rattled as it began its descent, and it seemed like a bug next to an American Airlines jumbo jet landing on the parallel runway.

I held my breath, clenched my jaw, and gripped the bottom of my seat. The plane aligned with the runway and its tires touched down with an abrupt *thump*! We were safely on land, and I let out a long sigh as Cesare taxied off the runway.

I was happy to be in San Jose as we deboarded and stretched our legs. Then Cesare and I picnicked on BLT sandwiches and lemonades while an airport attendant fueled up the Cessna for the return flight. Before long, we were up in the clouds again, headed back home. This time, we were buffeted by heavy turbulence. Air pockets caused the plane to lurch sideways and bounce up and down as if on an aerial roller coaster. Rocking and rolling.

"That's Magic Mountain down there," Cesare said, referring to the Six Flags amusement park in Valencia. I looked down and saw the roller coasters. He said, "Not to worry. It's just a little bumpy. Kinda like driving on a gravel road instead of a freeway." Then he grinned. "You know, using terms you can relate to."

My nausea returned, and for some reason I thought about an article I'd read in which turbulence had been compared to the hot wax bubbles in a lava lamp. The article described the bumps and dips as a result of the plane passing through the bubbles of warm air. I tried to picture my red lava lamp, hoping it would take my mind off my upset stomach.

I looked over at the gas tank gauge, and I'm no pilot, but the gauge looked awfully like the one in my Acura Integra. The arrow was bouncing just above the red range. Then I wondered if maneuvering through turbulence called for more fuel than a plane used in calm skies. I pointed to the gas gauge and looked at my brother. "Uh, Red Baron, aren't we running a little low on gas over there?"

"We actually have two gas tanks, so don't worry. We have plenty of fuel to make it back to Van Nuys. But if it makes you feel better, we'll fuel up in Burbank."

I peeked over and located the second gas gauge, which also appeared low, although outside the red range. "I never let my gas tank go below a quarter tank, so yeah, topping off in Burbank would make me feel better. Thanks."

Cesare pointed to the Golden State Freeway below. "Worst-case scenario, we could always land on the freeway."

My jaw dropped open, and I peered out the window at the freeway, which, at our altitude, appeared to be crawling with ants. "Land on the freeway?! Are you crazy?"

"No, it happens all the time."

"It does *not* happen all the time. I would have heard about it." I looked down again at the gridlocked traffic below and felt a strong urge to smack Cesare upside the head.

"I'm just saying that in an emergency we'd land on the freeway before crashing into those mountains over there."

"Okay, somehow hearing those words from the captain's mouth does *not* give me a warm and fuzzy feeling! I hope you work on your communication skills before you join the airline . . . schmuck!" I looked over toward the mountains and recalled the movie *Alive*, in which a plane full of rugby players crashed in the Andes mountains, and the survivors resorted to cannibalism. I stared at Cesare.

"What?" he asked.

"I was just thinking . . . if we crashed in the mountains, which part of you would I eat first?"

Cesare let out his distinctive howling laugh.

There was no need to land on any freeway. Cesare successfully completed his cross-country flight and earned his commercial certification. It was an extraordinary day. Being in a plane that small felt like an adventure, and my kid brother as the pilot made it unforgettable. Up there in the vast blue sky, we bonded.

Freddie reached out and wrapped an arm around me. "Ready to head home?" I nodded and we walked back to our car hand in hand.

As we weaved through traffic on the drive home, I thought about all the years I'd witnessed how much Cesare loved flying. I could tell that in the air there was a sense of freedom he'd found nowhere else. But I couldn't help wondering, what makes someone want to be a pilot? What kind of temperament does it take to pursue aviation as a profession? To potentially put themselves in the position of hero or villain. Looking out the window at a speck of a jet flying several miles over the surface of the earth, I pondered what it was about being airborne that offered such fascination? What compelled someone to take command of an aircraft and soar through the heavens?

Not satisfied with what solid land had to offer, Cesare had reached for the skies in search of new adventures, traversing over oceans and arriving at thrilling places clear across the globe. Up in the air, there was always a vast expanse of openness before him. As far as his eyes could see. Nothing else.

Back home, I sat at the desk in our office and thought about the many restrictions and obstacles Freddie and I had encountered during the drive home. Heavy traffic. Speed limits. Traffic signals. All meant to control the activity on the roadways. Wasn't this what Cesare tried to escape by flying? I knew the irony of it all, though: The sense of freedom Cesare often touted was misleading, because aviation is one of the most regulated businesses there is. Pilots need clearance for most of their actions. Takeoffs, landings, and ascending or descending to varying altitudes are all steps that must be approved by someone in a tower. Working for an airline, pilots have to abide by the rules and policies of the organization. In addition to regulations established by national aviation authorities such as the FAA, pilots working for international airlines like TACA also have to comply with the ICAO global standards. All these rules and regulations serve as guardrails that limit a pilot's autonomy, so the idea of freedom is relative. Pilots' wings are further clipped as they juggle priorities established by the airlines who want an edge over their competition. Priorities such as safety, customer satisfaction, efficient operations, and profitability sometimes further limit pilots from exercising true autonomy because they're bound by these goals. At any time, these priorities may be in conflict with each other, as when safety is compromised by rewarding employees for increased productivity when that productivity calls for risk-taking.

In fact, when it's unclear to an airline employee whether to prioritize, say, customer satisfaction or efficient operations, the onus falls on the worker to use their judgment. Absent clearly spelled-out policies, employees are forced to make decisions in the moment, which means the organization deflects responsibility to an individual. Then if things don't end well, it's the individual to blame, not the organization that created the policy gap in the first place.

I'd experienced this kind of conundrum firsthand. In an effort to fast-track and increase housing development in Los Angeles County, I was often asked to accept unorthodox construction practices or uncertainties in designs to get things done quickly. "Look at the big picture," I was told. "It's not that big a deal. Leadership is about taking risks." I had to let some things go that I thought were important, such as the need to determine long-term impacts due to a lack of current data. And I sometimes ignored the shaky feeling in my gut when I had been swayed to do something that hadn't felt right to me. These weren't life-and-death decisions like the kinds pilots must consider, but many times I'd been nagged by a lingering feeling that I'd been pushed to accept too much risk. The idea that during my career I might have put people at risk haunts me.

Risk-taking is alive and well in the airline industry. Cesare had probably landed on the Toncontin runway before under conditions very similar to those he faced the day of the crash, and I'm confident many other pilots did the same. In so doing, these pilots pushed boundaries, which increased risk. With no admonishment from TACA, this pushing of limits was deemed to be acceptable behavior, so it continued.

It was corporate culture's justification of great risk in the interest of great reward that explained why TACA Airlines chose to continue to use Toncontin Airport despite its flawed infrastructure even under the best weather conditions. The fact that TACA was willing to tolerate the risk factors of a tropical storm and complaints of hydroplaning reveals their culture of high-risk tolerance. That choice in itself was a message to its employees about acceptable risk. Had safety been the company's priority, flights would have been delayed or diverted. But they weren't because profit was more important than protecting human life.

I've also given a lot of thought to the responsibility of consumers in this tangled web of risk versus reward. Passengers can be very vocal when their plans are interrupted. People don't want to tolerate delays because their loved ones are waiting to pick them up at their destinations. A missed connection can cause us to arrive late to an important business meeting, or it might cut our dream vacation short. And we complain because we know how a delayed flight disturbs our plans on the other end, but on the flip side, we don't know what tragedy might have just been prevented by a decision-maker who erred on the side of caution. Because we don't possess magical foresight, we don't know what kind of disaster a conservative decision might have just spared us, so we don't feel gratitude for the decision.

Pilots face competing priorities and are pressured to deliver on-time performance and excellent customer service. This may be in conflict with the decision to land a plane at one airport instead of another. In fact, as Cesare was near landing and was weighing his options, he asked TACA operations whether he should continue to an alternate airport. Recognizing the lack of clearly defined protocol under these circumstances, he tried to place the decision on TACA. Perhaps it was Cesare's way of forcing accountability from TACA leadership.

Did Cesare face competing pressures on that day? Did they ring in his ear? *Why* did he choose to land the damned plane? Why not circle around again? Why not turn back? Why not just reject thoughts of the possible fallout from airline executives or demanding passengers? Why not make the unpopular choice?

Brother, why didn't you tell them all to go to hell and err on the side of caution? *Why? Why? Why?*

I sat at my desk and lashed out at whatever was within reach—in this case, it happened to be a wall, the entryway wall to be precise. With a guttural scream, I slammed my fist into the wall. I wanted so much to puncture the drywall, to feel my hand crack through to the other side, to rip through the structure and watch flakes of plaster rain down on me. Instead, there was a loud thump as my fist made contact. Then pain pierced my hand and traveled into and up my arm. I crumpled to the ground, tears of pain and frustration collecting like a pool around my sorry self. "Goddamn you!" I yelled to no one.

As an engineer, I should have checked where the studs were before slamming any part of my body against a wall. The stupidity behind the injury I'd just caused myself pissed me off even more.

I sat in silence at the base of the stairs that led to the second story of our house, holding my throbbing hand in my lap. My mind was a swirl of anger, anger at Cesare, at Toncontin, at every regulating agency that hadn't prevented my brother's crash. Then suddenly I had a jolt of clarity.

Was I most angry at myself?

Angry at myself? *Was I having an epiphany?*

My father had entrusted me with a mighty mission: *Protect your little brother. Always protect your little brother.* He believed I had it in me to accomplish that gargantuan task, that I was worthy of the superhuman undertaking. My father had tapped *me* to carry on protecting Cesare when he no longer could. And I'd accepted the mission. So as Cesare's life had progressed happily and without injury, I'd grown up believing I'd been successful in shielding him from harm. I'd been given the assignment by my father, and for years I believed my brother's charmed life had somehow been my doing. He'd made thousands of landings at regulation airports and risky ones, too, and somehow, I took partial credit for that. With each passing year, I grew more confident Cesare was being kept safe and that I was doing a great job of honoring my father's wish.

Until my little brother was killed.

There it was, a new awareness of what was so awful about this catastrophe beyond what other people saw. Steel mangled, people dead, families broken—this is what other people saw and suffered with. But for me, there was another layer of pain attached to the disaster. It was the awareness I'd fallen short. As I sat crumpled and defeated, I realized I was now heartsick not only because my brother was gone but because his death was glaring proof I'd failed him.

10

Postmortem

About three months after the accident, our family received the autopsy report from the American embassy in Honduras. I finally had the specifics of Cesare's death, but I walked around with this forensic investigation in my briefcase for weeks before finding the courage to read it. Then one night, I decided I was ready to learn the truth, so I sat at our dining room table alone with the report staring back at me. In keeping with my training as an engineer, I equipped myself with the materials I thought I'd need to handle this most consequential task: a medical dictionary, pictures of a human skeleton, and a red felt-tip pen.

I opened the report and scanned midway down the first page. There, in big bold capital letters, was the answer I'd been waiting for: *CAUSE OF DEATH: CARDIAC CONTUSION.* Online I learned this was defined as bruising of the heart due to blunt-force trauma. *Bruising.* That sounded so benign. Certainly not deadly. I read that 50 percent of blunt chest trauma cases are a result of motor vehicle collisions and that 20 percent of these types of collisions result in deaths. If the contusion occurs because of enough force, the heart can be compressed between the sternum and spine. Often, these types of injuries are not survivable, and patients typically die at the site of the incident.

As an engineer, I was very familiar with the laws of physics. Specifically Newton's Law, which states that a body in motion remains in motion until an outside force acts upon it. In high-speed accidents such as Cesare's, there are actually three impacts. The initial impact was the obvious one—the plane came to an abrupt stop as it crashed against the embankment. The momentum caused the fuselage to buckle and bend out of alignment. According to the preliminary investigation report released days after the crash, the flight data recorder indicated the plane was traveling at fifty-four knots, or about sixty-five miles per hour, before impact. As the plane hit the embankment and came to a violent stop, the human bodies inside the airframe continued moving forward until they came in contact with "an outside force." That must have led to people crashing into the backs of the seats in front of them and maybe even into each other.

Then came the second impact. As the cockpit buckled, Cesare's body must have crashed against either the seat restraint or the cockpit controls or both. His fifth thoracic vertebrae was broken, and a deep head wound exposed his cranium. He also sustained severe neck trauma, and multiple fractures and breaks of his feet, legs, and pelvis. Some of his bones ripped through his soft tissue, tearing nerves, arteries, and ligaments, causing severe hemorrhaging.

The third impact delivered the final blow. Although Cesare's body also came to a violent stop, his internal organs continued moving forward, smashing against other body parts—the heart striking the sternum; the brain hitting the skull, causing a cerebral edema; and the lungs assaulted by ribs. His internal bleeding was extensive. The report said his chest and abdomen suffered the brunt of the impact, which caused extensive damage to his ribs and heart, and the force resulted in a traumatic rupture of the thoracic aorta, the body's main artery, which resulted in instantaneous death. I paused for a moment to catch my breath and control my shaking hands.

The autopsy report was telling me Cesare's death had been immediate. I tried to calm myself with a reminder that an immediate death would have been merciful, that I should take comfort in knowing my brother didn't die slowly in great pain. Then, with all the magical thinking I could summon, I wished for an alternate ending to the story. What if he'd survived and come home alive? What if he'd walked away from that crash? I knew if Cesare had lived, he could be happy only if everyone else on that plane had also survived. If he'd lived while others had died on his watch, he would have carried crushing survivor's guilt for the rest of his life. And what if his injuries had prevented him from flying again? Maybe tethered to a bed or a wheelchair. His wings clipped. Doomed to life on land. It would have been his idea of hell on earth. His spirit would have been broken. One way or another, the crash would have left him dead.

Thanks to the information and data before me, the many reports I had reviewed, and the calculations I'd run, the pieces of the puzzle came together. I now knew what happened to my brother's body. But nothing could tell me what had been going on in Cesare's mind. I knew how well-trained, how well-prepared my brother was to land safely in all kinds of conditions. But what goes through a pilot's mind as he sees his plane is about to careen into an embankment? How fast can a mind process what's happening and what might be done to prevent or mitigate it? Did Cesare know he was going to die? And if he did, what were his final thoughts?

I suppose I'll never get those answers. As an engineer, sometimes the best I can do is develop a sound hypothesis, and my gut told me my brother had no idea the end was near. I chose to take comfort in that, too.

It's been said that people are most alive when they're in love, when their sensations are heightened by this kind of connection with another person. Cesare's true love, the thing that had always made his heart soar, was flight. It was the constant in his life that let him be exactly who he was meant to be and maybe the only thing that ever really captured his heart. Cesare would have done anything for this love. And he did, through the most selfless of acts. He gave it all.

Several hours after I read through the autopsy report, I was back on a plane headed to El Salvador. TACA had requested a meeting with the family to brief us on the preliminary results of the accident investigation, so I secured an aisle seat on a red-eye. A few minutes after I'd taken my seat, an older gentleman politely pointed to the empty seat next to me, and I stood to let him in. Once he was settled, seat belt buckled, he turned to me, extended his hand, and introduced himself. I shook his hand. "Nice to meet you, too. My name is Rossana D'Antonio."

He stared at me and repeated softly, "D'Antonio." He'd figured it out. He knew. I gave him a half-smile and nodded.

"I'm so sorry," he said.

"Thank you." I slipped my headphones on, hoping to avoid conversation. It worked. The gentleman pulled out a book from his carry-on, *Love in the Time of Cholera.*

The plane took off, and shortly the in-flight entertainment was announced, but I wasn't really listening. Ever since going through the autopsy report, my thoughts had been haunted by images of Cesare's body. The ruptured heart. The broken vertebrae. Ribs puncturing lungs. The brain smashing against the skull.

Exhaustion finally sent me into a restless sleep.

I stirred in my seat and opened my eyes to unfamiliar surroundings. Where was I? A metallic chamber. A cockpit. My gaze fixed on the thick blanket of fog outside the window. The cockpit windshield. What was I doing there? Slowly, the dirt embankment came into focus, and an eerie quiet gave way to a muffled mix of buckling metal and hollow echoes. Was I underwater? In a womb? Being born? Or dying?

Fine dust particles floated through the air, and I was surrounded by clutter, strange, nonsensical clutter. Loose sheets of paper were strewn over the eerily dormant control panel. Eyeglass frames without lenses sat ominously out of place on the floor near my feet, peering at me, a hollow stare. A loose black shoe lay upside down on the opposite side of the compartment, abandoned by its mate. My harness, locked taut, pinned me in place, pressing on my chest. I whipped my head to the left, and there was Cesare, slouched forward, strapped in his seat.

What was I doing here?

His left arm hung limply, and I watched the trickling of blood near the black leather band of his Cartier watch. I called out, "The blood will stain the leather!" But did I? My mouth hadn't moved. Cesare was completely still. I watched his chest, desperate to see the rise and fall. But there was no movement.

I shouted, "Hey, are you awake?" I reached for him, my arm heavy and sluggish. The harness locked me in place, bullied my chest, and I struggled to breathe.

"Wake up!"

How was it I could hear my words, but my mouth wasn't moving? My heart crashed against the inside of my chest. Wake up! I'm here! *A booming echo screamed at me,* protect your little brother. Always protect your little brother.

As I strained to touch the four gold bars on his shoulder, the space between us became distorted and warped. Cesare's body seemed to retreat. Where was he going? The more I stretched toward him, the greater the distance between us grew. Was the cockpit growing wider? No! Wait! Cesare! Where are you going? Stop! *He was disappearing! How was he fading? Where did this fog come from?* Cesare, wake up!

Then suddenly I was being suctioned at supersonic speed into a dark tunnel. I squinted, desperate to stay focused on Cesare's fading body. "No!" I screamed. "Don't go! Wake up!"

I jolted awake, and the nice gentleman next to me looked over with kindness in his eyes. He patted my hand that lay on the handrest. "It's okay. It's going to be okay," he said.

Was he telling me it had only been a nightmare and I was okay now that I was awake? Or were his soothing words intended to be bigger, letting me know that all would truly be okay despite this disturbed reality I was living? I wanted so much to believe the latter as I felt the wheels of our plane touch down.

11

False Truths

I stared out the window of the taxi shortly after arriving once again in El Salvador. As we pulled away from the airport, I was greeted by ashen skies and gray-white clouds that showered thick sheets of precipitation, torrents of water that thundered on car hoods and splattered the street clean. Thick raindrops raced each other in long streaks down the windowpane. The humidity was stifling and draped a thin layer of stickiness over my skin.

El Salvador. The land of the Mayan Civilization. At just over 8,000 square miles, El Salvador is about the size of New Jersey. It's also the smallest country in continental America. As a kid, I sat for hours studying the multicolored *World Atlas*, carefully pinpointing my mother's homeland. "El Salvador is the *pulgarcito de America*, the Tom Thumb of the Americas," Mom would say, "because it is the smallest of all five countries in Central America." I loved thinking of my beloved El Salvador this way and played at assigning the remaining four Central American nations a corresponding finger on my hand. I was proud of El Salvador, "The Savior." But the reality for the Salvadoran people was one of political instability and socioeconomic inequality. Passing the lush, green mountains, I noted the little aluminum shacks with rustic corrugated roofs that peppered the mountainside.

In California, only the very rich can live on the hillsides. They spend millions on deep foundations, so their homes remain stable. I know; I spent a large part of my career designing them. But these hillsides were inhabited by the poor and needy. It was their shacks notched on the side of the mountain and they who risked their homes to unstable ground, to heavy rains washing the soil out from underneath the makeshift structures and creating rivers of mud that sometimes washed away entire villages.

Along the hillside, I saw the unkempt tropical overgrowth interrupted every so often by a walking trail where women carrying jugs of water made their way home. Slouching concrete poles that lined the highway carried electricity from town-to-town, and garbage was strewn all over the side of the road. Lazy skeleton dogs with sad eyes lay along the ground and lifted their heads as the taxi drove by. Crushing poverty had been El Salvador's reality for too long. The have-nots far outnumbered the haves, a cruel social inequity resulting from centuries of systemic injustices.

During El Salvador's civil war, it was this social inequity that drove hordes of Salvadorans to the US. At that time, the evening news showed videos of frightened people crossing the Rio Grande, balancing baskets and bags of their most precious belongings on their heads in a desperate attempt to get to the Promised Land. Many a night I watched images of the floodlights guiding the border patrol guards in their jeeps as they raced after dozens of people scrambling in different directions like ants. Newspaper articles told stories of "coyotes" who charged thousands of dollars to escort their victims, like cattle, in overcrowded trucks but often left them to die or abandoned them partway through the journey to cross the Chihuahuan Desert on foot. Countless desperate El Salvadorans braved death in hope of escaping the civil war at home and making a better life in America. America the beautiful.

The taxi pulled up outside Mom's house hidden behind the tall brick wall Mom had commissioned during El Salvador's civil war. "One can never be too safe," she'd said.

It had been a little over three months since I'd been here, and although I was happy to be back, I was uneasy. The next day, I would meet with TACA representatives to hear what they knew so far about the cause of the crash. I handed over several bills to the taxi driver and thanked him as he set my bag by the door. No sooner had I rung the doorbell when Mom flung the door open and practically fell into my arms. "I've missed you," she said, her voice cracking. "Come in. Your sister's here."

"Hey, you," Mars said as I entered the house. Then she hugged me. "You're a sight for sore eyes." Her words were more than a cliché. My sister looked exhausted.

I usually flew to El Salvador on a red-eye, and I could always count on Mom to greet me with brunch, her favorite meal of the day. The table was set for three as we sat down to enjoy *pupusas*, El Salvador's national dish—griddled corn cakes stuffed with cheese, beans, and pork and usually eaten with a pickled cabbage relish. Urban legend was that it was a sin to cut pupusas with a knife because they were considered to be made with a godly grain, so pupusas are traditionally eaten by hand. I could never get used to the messy tradition, and whenever I'd ask for silverware, I was mocked by the locals, "Sacrilegious!" The only thing better than pupusas were the side dishes of fried plantain bananas and refried beans. All of it was sinfully delectable.

"So about tomorrow," I said as I sipped my piping hot coffee, "Mom, I'd prefer you not go to this meeting. Mars and I can fill you in later." I didn't expect to hear any resistance. Except for one meeting at TACA headquarters a few days after the accident, Mom had avoided discussions of the crash. It was just too painful.

Mom said, "Drop me off at the chapel at the cemetery compound, and I'll visit with your brother." Every Sunday afternoon without fail, Mom went to the cemetery. During the rest of the week if she happened to be in the neighborhood, she squeezed in an extra visit. Unlike my mother, I found no comfort in standing before a grassy plot of land under which a loved one had been laid to rest. I didn't think it was Cesare under that patch of sod, and

I'd always found disposing of a human body in the ground disturbing. The thought of the chemical decomposition of a human body increasing the moisture in a confined space and creating a perfect environment for microorganisms to flourish and feed off the physical remains of a loved one was just gross to me. In fact, shortly before Cesare's burial, I tried to convince Mom to cremate his remains. "Mom, that's what he would want. Do you remember him saying so?"

She'd snapped, "I refuse to destroy my only son's body!"

And that was the end of the discussion. Months later, I found comfort in knowing her visits to the cemetery gave her so much relief. At the cemetery, she felt close to her son.

I had a bad feeling about the TACA meeting ahead. My gut was telling me the airline might follow through in making Cesare the scapegoat. Freddie often accused me of creating negative future fantasies to torment myself with what-ifs. He said it was wasted emotion and energy because things usually worked themselves out. As much as I loved my husband's optimism, my career had taught me that things sometimes, even often, go tragically wrong.

The next day, Mars and I waited in the lobby of the TACA headquarters building located along the foothills of Santa Elena, an exclusive neighborhood in El Salvador's capital city. Out the window of the multistory building, I saw a thriving urban community that had retained its natural beauty among multistory residential and modern commercial buildings. Santa Elena showcased how much progress El Salvador had made since the end of the civil war sixteen years earlier. It was home to the American embassy, financial centers, and several US-based companies.

I stared out the wall-to-wall windows toward the American embassy located just catty-corner to the TACA headquarters. I'd read somewhere it was the largest embassy in Latin America, with the highest level of security, and that it had been the most expensive to build in the region. It had taken years for the U.S. State Department to get enough federal funds to fulfill the top security needs due to the region's civil unrest during the 1980s. But after multiple delays, construction was finally completed at the tail end of El Salvador's civil war, just in time for the historical signing of the peace treaty that vowed to ensure a true democracy. The red, white, and blue United States flag waved in the wind as it towered over the high-security gates and concrete walls. But the massive American embassy resided in a now-peaceful country, so the fortress seemed like overkill.

In the lobby, Mars and I approached the TACA reception desk. Behind the desk sat an ebony-haired young lady wearing a gray blazer and a silk neck scarf that matched the bright red of her lips. Mars spoke first. "We are the family of Captain Cesare D'Antonio, and we have a three o'clock meeting."

The young woman flashed a big smile that showed her pearly white teeth. And what was it I saw in her eyes? Was it hope? Her expression was considerably different than the forced smiles and downcast eyes of months ago, all the expressions of sympathy. So many people feeling sorry for us. I much preferred this greeting. Because despite heading into a meeting where I would learn more about the cause of the crash, I was hopeful, optimistic I might hear Cesare hadn't died in vain. Maybe TACA was going to announce some new safety policy created in response to the crash. Maybe Mars and I were about to hear their proposal for airport improvements.

It had been only five months since the accident, but already I'd learned of changes within the airline. Changes I interpreted as TACA moving on. For starters, there was no longer a Flight 390 on their schedule. The route from San Salvador to Miami with stops in Tegucigalpa and San Pedro Sula still existed, but the flight number had been retired.

My unbiased self told me that of course they'd try to put it behind them quickly. Who would want to call attention to a deadly plane crash? Planes skidding off runways and smashing into embankments was bad for business. Easy to understand that. But as Cesare's sister, I felt history and my brother were being erased and TACA was scrubbing the whiteboard fast.

A new contemporary logo of an abstract red parrot now adorned TACA's planes and billboards. Changing a logo is a huge deal; it has to be replaced on planes, paper, uniforms, merchandise, billboards—everywhere. A transformation like that could cost hundreds of thousands of dollars. Maybe more. But it was practically a requirement for TACA as they covered up the past and embarked on a fresh start. On the flight over, I'd noticed the crew now wearing charcoal-gray uniforms with light gray bands on the sleeves. Bygone already were the black uniforms with gold stripes on the jacket cuffs; the uniform Cesare had worn was already a thing of the past. Since I'd never seen Cesare in a gray uniform, it wasn't likely I'd ever again experience the momentary jolt, the wild imaginings, the tricks my mind had always played when I saw a TACA pilot walking toward me.

TACA's corporate renaissance was a brilliant scheme to rebrand the airline. Like Cesare's family and friends, employees of the airline needed to recover, and what better way than to start with the cleanest slate possible? I'd seen other businesses roll out scrub campaigns. Corporations focused solely on their bottom lines willing to whitewash truth, some even deliberately venturing into cover-ups. But for me this campaign was personal.

Marta Liliam, who was not only our liaison to the airline but someone who'd earned our trust, greeted us in the lobby with strong hugs. In the past five months, she'd been the one my family interacted with on all matters related to Cesare. Marta Liliam was TACA's messenger, but she had a way of delivering messages with humanity, kindness, and empathy. She knew the soothing value of a touch or a warm embrace. She might have been the one who gave me hope TACA might honor my brother appropriately.

Mars and I followed her into a conference room where two men stood next to a small conference table. What struck me right away was that neither of the men were wearing business suits. No ties. No blazers. I was slightly offended by this level of informality. What, these guys couldn't be bothered to dress up for my family? To me, it said Cesare's memory meant nothing. They had already moved on.

The taller of the two extended his hand. "Manuel Antonio Mojica, flight safety director. It's a pleasure to meet you. I am so sorry for your loss."

I'm sorry for your loss. How many times in the previous few months had I heard those words? How many times had I heard them throughout my life? What an annoyingly overused expression. *Why* couldn't people come up with their own damned sentence to express sympathy? Was *one* original sentence too much to expect in response to the end of a life? Mojica stood there in dark slacks and navy-blue polo shirt waiting for a reply, but I had none for his hollow words that sounded tinged with insincerity. I nodded and offered a pinched smile.

The second man introduced himself as the director of operations, but I didn't catch his name. Mojica gestured for us all to have a seat, and I sat right next to him rather than on the other side of the table. I wanted to avoid creating an "us against them" vibe. And I wanted Mojica to feel my presence. Marta Liliam sat next to Mars, across from me.

Mojica began, "We're here to present the preliminary information from the investigation." He explained they had another meeting scheduled after ours, so following a few pleasantries, the lights were dimmed, and we stared at the screen on the wall where Mojica projected images and data in a PowerPoint presentation.

Right away I asked, "Can I get a copy of this presentation to take with me?"

Mojica looked away from the screen and said, "I'm sorry, but this is still preliminary information. We can't give out copies."

I looked directly into his eyes and responded, "That's okay. I can follow along." I could and I did.

Mojica breezed through the slides as if he was in a rush. *What's the rush?* I wondered. Were we not worthy of more than an hour to TACA, the company our brother flew thousands of people safely for, and then gave his life for? Then it hit me. The problem was *gave his life for.* That's why Mojica was speeding through this process. He was here to deliver a message that didn't fully acknowledge, much less honor, the sacrifice my brother had made, so he wanted to get out of here as fast as he could.

The director of operations, a short, pudgy man who wore wire-rimmed glasses on his round face, said little, except to occasionally interject and echo what Mojica was saying—to reinforce a point, I supposed. I found his parroting awfully annoying. Mojica and his Echo. Reminded me of Laurel and Hardy, Gilligan and the Skipper, Don Quixote and Sancho Panza. All of them pairs that included at least one clown.

The figures and data felt like pellets being shot at us. And as much as I tried to dodge them, each word singed my face. My hand flew furiously across the pages of my journal as I took notes. I was determined to absorb every single piece of information being thrown my way. As an engineer, I understood the mechanics behind the data being presented. And the facts Mojica was presenting were very familiar to me by now; I'd been researching and studying the crash for months. But trying to keep up with the presentation while also trying to scribble and process all his information was rough. It was like practicing tennis ground strokes against a ball machine that sent shots deep to the baseline corners at full speed. My head snapped back and forth from the screen on the wall to my notes on the table.

TACA Airlines Airbus A320
Flight 390
Irish Registry IE-TAF
135 passengers and crew
Wind information 190 degrees / 10 knots
Wet runway
Aircraft landing weight 63.5 tons
Maximum landing weight 64.5 tons
Aircraft cleared to land on runway 02
Autopilot OFF
Flight directors OFF
Indicated air speed at touchdown 139 knots
Ground speed 159 knots
Tailwind 10 knots
Runway elevation 3,297 feet
Displaced threshold 700 feet
Landing distance available 5,410 feet
Touchdown at 1312 feet from displaced threshold
MAX REV
Nosewheel touch down 7 s after main landing gear (MLG)
Manual braking 14 s
Maximum pedal braking 21 s
70 knots
IDLE REV
257 feet stopway distance
24 s
54 knots
65-foot embankment
26 s
End of report

Then the screen went white. I took a deep breath. Had I been holding my breath all along? Mojica looked at me as if reading my mind. "There have been many questions about the runway that was cleared for landing. Flight 390 was initially cleared to land on Runway 20 by the control tower, but Captain D'Antonio asked to be cleared to land on Runway 02 because of the fog. As the captain, it was his prerogative to land on an alternate runway. He was granted permission to land on 02 with caution due to wet runway."

Mojica's stumpy sidekick nodded quickly.

Mojica added, "I would say that it was excess confidence."

My head snapped back. "*Excuse* me?"

He kept his eyes locked on mine. "Excess confidence."

"I'm sorry, are you suggesting he shouldn't have landed?" My neck pulsed furiously.

Mojica cleared his throat. "Given the circumstances, it is my opinion that he should have continued to the next airport."

I said, "Given the circumstances of what you know *today*. But on *that* day, he was cleared to land." Someone's chair squeaked. I didn't take my eyes off Mojica.

He straightened his pen so it sat parallel to a notebook on the table in front of him. "A pilot is in command of his aircraft and ultimately can overrule the control tower."

I looked across the table at Mars. She was looking down at her lap. This clash was going to be all mine. "But there was *no communication* from the control tower that indicated he should have overruled them. He was cleared to land!"

Silence.

I kept pushing. "Would there have been any negative repercussions if Cesare had decided to turn back or land in an alternate airport?"

"None, absolutely not!" Mojica snapped.

My eyes locked on Mojica's, squinting slightly as if that would give me a clearer picture of what he was truly saying. I didn't believe him. "What about the recent repaving of the runway? There's no evidence of drainage grooves. Pilots have complained about hydroplaning."

Mojica took a long breath and softened his voice. "Look, Toncontin Airport is a difficult airport to maneuver, no question about it. We provide supplemental training to our pilots, and additional certifications are required to land there, precisely because of the nature of the airport. But in the end, it is each pilot's decision to pursue the training required to fly in and out of Toncontin. Captain D'Antonio chose to get the extra training. He understood the risks of the airport, and he agreed to fly in and out of it anyway." He pushed the notebook on the table away from him. "I personally decided years ago not to pursue that certification. Flying into Toncontin was too stressful."

Sancho Panza nodded like a dashboard bobblehead. I realized then that Mojica was a pilot himself, a pilot who just said out loud that for him, the conditions at Toncontin Airport were . . . what? Too risky? Too scary? Too potentially deadly and therefore not worth facing as a pilot? This top executive, *TACA's flight safety director*, had just acknowledged that he personally refused to take on the high risk that came with flying into Toncontin. But that same flight safety director had no problem subjecting other pilots and countless passengers to a dangerous environment.

Silence settled over the room, but inside my head the words blared. *Excess confidence* with a dash of *you knew what you were getting yourself into*. This was the airline's position. Not a shred of airline responsibility. They were saying, "Toncontin might be a dangerous airport, but you don't have to land there if you don't want to." I looked over at Mars. Her face was pinched into a grimace. She was in pain and holding back tears.

"The investigation isn't over," I said, pointing out the obvious.

"No, it's far from over," Mojica said.

"Then how can you have *concluded*, 'excess confidence'?"

Mojica stared blankly at me. "This is our conclusion at this preliminary stage."

"How much longer for the investigation to be completed?"

"A few months."

"We want a copy of the final report."

Mojica nodded and began to gather his things.

TACA Airlines would accept no responsibility for choosing to fly in and out of a dangerous airport. No one would assume responsibility for Cesare's having been cleared to land. There hadn't been a word about what it meant that Cesare hadn't been advised by anyone in the control tower not to proceed with the landing. They were trying to sell the idea of excess confidence. But none of the facts pointed to this conclusion. The facts pointed to a potentially flawed runway.

And what was excess confidence supposed to imply, anyway? Don't all pilots have excess confidence? Doesn't the job call for it? Who else chooses a career that involves being in control of a 150,000-pound flying machine and rises to 35,000 feet over the surface of the earth so they can transport hundreds of people from point A to point B if not those with bonus confidence? Don't we want excess confidence in our airline pilots, neurosurgeons, and members of a bomb squad? And now Cesare was being accused of having too much of the quality that got him the fucking job!

Mojica's voice, tinged with impatience, interrupted my thoughts. "If there are no more questions, we have another meeting to go to." He rose and began to pack his PowerPoint projector, fumbling with the power cord. I sat stunned but managed to clamp down on my anger long enough to firmly shake the hand he thrust in my direction. Then he left the conference room with Sancho trailing behind.

As I watched them walk away, I thought of a quote I'd read by the renowned safety expert Sidney Dekker, *Cause is not something you find. Cause is something you construct.* I could certainly attest to this. In 2018, our field crews were using handheld radios when the Woolsey Wildfire raged through the Santa Monica Mountains, and we lost communication. A colleague, concerned for their safety, tapped into an emergency fund to purchase several satellite phones for our crews in order to reestablish communication. But an executive manager, sitting in the comfort of his office, scolded my peer, as if reprimanding a child, for buying too many phones. My colleague's "crime" was a purchase interpreted as "excessive without proper authorization." Never mind that my coworker had kept our staff safe with reliable communication equipment during a natural disaster. Knowing that an after-action report would be prepared documenting major decisions, including financial decisions made during the emergency, the executive manager chose to deliberately construct a cause that laid blame on his subordinate for what he deemed an excessive purchase. By doing so, he deflected attention from himself and the failure of our organization to implement a policy that established clear expectations during emergencies.

Excess confidence was the cause TACA had constructed. And it stood on as much shaky ground as the deadly Toncontin runway itself.

My hands were numb and my legs shaky. Mars, Marta Liliam, and I remained seated at the table in awkward silence. Three women. The air felt less oppressive now that the men had left the room. Many times throughout my career, I'd found solidarity in female colleagues who shared with me not the highs of victory but the moments that seemed most bleak and hopeless. It was this shared despair that sparked the strongest levels of resolve in me. It's what I felt as I looked at Mars and Marta Liliam.

"Would you like some water?" Marta Liliam asked, breaking the silence. Although my mouth felt full of cotton balls, I politely declined. We thanked her for everything. Despite her position with the airline, Marta Liliam had stood in solidarity with our family and had graciously brought humanity to all our business dealings with the airline. Her warmth and caring eased our burden, and it was now time to say goodbye to her. Mars and I each quietly hugged Marta Liliam, then we turned to go.

We left the building, got into our car, and began the drive to the cemetery. Misty rain dotted the windshield and turned the road slick. During the ride, my sister and I didn't speak. Mojica's voice snaked in and out of my thoughts but was soon interrupted by *take care of your little brother. Always take care of your little brother.*

12

Blame Game

Later that night, I sat alone in Mom's living room sipping a glass of pinot noir and scanning my mother's décor. Every horizontal surface of my mother's house was topped with family photos. I'd always been critical of the clutter, but on that night, I was comforted by all the reminders, all the images of days long gone when my worries had been no bigger than a squabble over who'd eaten the last piece of cake.

Mom hadn't asked about the meeting, so I was free to process the day's events on my own. Why had I been surprised the preliminary investigation had concluded pilot error? Had I really expected anything different? Seventy percent of airline accidents are blamed on the pilot. It's easy to sit in an office reading a step-by-step depiction of what happened, including the outcome, and conclude that one person is responsible. It often serves multiple agendas to craft a tale that lays full blame on one individual, a villain. The person at the front line, the one who last touched the system, the complex system that is aviation. Perhaps the most dangerous byproduct of such targeted blame is that in blaming only one cause for such a disaster, those in positions of power shift attention away from the systemic problems that often lead to unsafe air travel.

What about the built environment? What about the runway infrastructure itself? What about the safety concerns TACA pilots had expressed about Toncontin Airport? What about the other nine accidents that occurred there over forty-six years? And what about the fact that those accidents had happened in similar circumstances? Did *all that* not indicate Cesare's crash wasn't likely pilot error?

And what about the public acknowledgment by President Zelaya that there was a known problem with the airport? What about the recent complaints about hydroplaning? What about all that? Mojica had been defensive. That was a tell. A lot of Mojica's presentation compounded my already heightened skepticism.

Hindsight was everything. We now had a linear depiction of the events of that day. We had all the information. And while data was absolute, *how* we interpret the data could be prejudiced. Yes, a convenient cause had been constructed. In the aftermath of an event, we're all experts. We can all play judge and jury. But the crew didn't have the benefit of hindsight. Their awareness of the events came only as the events were in progress. Over those twenty-six seconds, the crew members assessed a flurry of data and acted on real-time information—facts provided by the control tower, situational awareness from the surrounding environment, and data from their instruments. Yes, Cesare knew the risks. He'd been a longtime employee of TACA Airlines. He'd chosen to pursue all the higher-level training and certification as established by the airline. And he'd successfully landed in Toncontin fifty-two times. This had been nowhere near his first rodeo. And now, TACA used Cesare's staying power to their advantage, dismissing the risks because Cesare remained willing to take them on.

Those investigating the crash looked at twenty-six seconds and one man's decisions. But Cesare's performance as pilot wasn't an isolated event. His behavior that day was a product of the organization he represented—the culture, the systems, TACA's level of acceptable risk. I have years of experience leading after-action assessments following emergency response operations, so I know a valid investigation should focus on a holistic review of not only the actions of the frontline workers but also the entire institution that created the workplace environment. I go in after an earthquake, a bridge collapse, a wildfire, or a flood and help determine what went wrong and what needs to happen for things to get back to going right. These assessments are meant to evaluate performance of frontline workers, analyze human response to key situations, review the organization's critical procedures and policies, highlight breakdowns in organizational defenses, illustrate lessons learned, and provide recommendations for improvement. I've conducted thousands of such investigations, and not once have I ever concluded human error. What I *have* concluded was that frontline workers in these scenarios were affected by poorly established critical procedures or a lack of clearly defined policies. In other words, a system failure.

My research into my brother's crash had already led me to believe pilot error wasn't the cause of the TACA Flight 390 accident. It was a symptom of a systemic problem of high-risk acceptance. There was rarely a single cause of an accident the caliber of my brother's plane crash. Instead, a series of cascading events were usually at fault. But TACA investigators ended their research after they landed on pilot error because that conclusion was tidy. And it was a very convenient conclusion when the person being blamed was no longer alive to defend his actions. Point the finger at the pilot because we're certainly not inclined to incriminate the airline's CEO, the director of the FAA, or the airport operator. Theirs are the faces that linger in the shadows, while the real face of the industry belongs to the crew. We know them. We see them lugging their carry-ons through airport breezeways. They stand in line alongside us at the terminal food court to buy grab-and-go meals before their flights. They're just like us, showing up and putting in another day of work. And while we're up in the air with them sharing our journey, we establish a relationship of sorts. They get us where we need to go and along the way make us feel safe and comfortable.

Blame is a powerful thing. It can help break down a frightening event into something more tangible. Who hasn't feared being a victim of a plane crash? How many scenarios do we ever put ourselves in that render us so out of control? We convince ourselves that pilots know what they're doing, which is how we trust we'll be safe when we fly. Then, if something goes wrong, we can point to the pilot, and this gives us the comfort of knowing who to blame.

Upon hearing of an aircraft incident, the media and its viewers usually speculate about pilot error, and in no time the theory spreads. Social psychologists refer to this phenomenon as blame conformity. We're easily influenced by a dominant voice that establishes the message. In the case of TACA 390, the dominant voices were the airline and the airport owner, which protect reputations by laying blame on one individual. And the public usually follows suit.

TACA landed on excess confidence as the basis for pilot error. Their conclusion was that Cesare never should have landed on that runway, and they took the position that because they now understood the crash was caused by just one individual acting in error, a disaster like that wasn't likely to happen again. This meant, once again, they could ignore the history of deadly accidents at Toncontin, hoping those of us who travel by plane would do the same. Blame the pilot, and everyone can get on with business as usual. Come fly the friendly skies. And profits will keep rolling in.

I wasn't sure what my plan should be after hearing TACA's report. I just knew I couldn't sit idle and do nothing. I knew there was more to the story.

Take care of your little brother. Always take care of your little brother.

13

Southwest Flight 1248

On December 8, 2005, Southwest Flight 1248 overran the runway while attempting to land at Chicago Midway International Airport, then crashed through an airport fence and slid across a road. I heard about the accident while I was Christmas shopping for Freddie's and my upcoming holiday trip to El Salvador. At the time, I was focused on my Christmas list, so after confirming Cesare was nowhere near Chicago, I only half listened to the news story. But many years later, when I was deep into researching my brother's crash, I downloaded the accident investigation report on my laptop at home and dug into details of Flight 1248.

The aircraft, a Boeing 737-700, originated its flight in Baltimore and was headed to Chicago, then to Salt Lake City and Las Vegas. Although snow had fallen on Chicago throughout the afternoon, snowplows had been clearing Midway's runways.

Chicago Midway is located about ten miles southwest of downtown Chicago at an elevation of about 620 feet in an area that includes land used for residential, commercial, and industrial purposes. It has several runways, but because of cloud cover that day, only runway 31C was operational. Runway 31C has a displaced threshold, so its usable landing distance is only 5,826 feet, which is over 400 feet longer than Toncontin's runway 02. And unlike the runway at Toncontin, runway 31C is constructed of grooved concrete rather than grooveless asphalt.

As the Boeing 737-700 was nearing the airport, air traffic instructed the pilot to enter a holding pattern while the runway was being cleared. Shortly thereafter, the control tower cleared Flight 1248 to land but warned the pilot to expect mixed braking-action conditions, by which they meant good braking action for the first half of the runway and poor conditions for the second half. When landing, the pilot applied the plane's brakes, but instead of deploying the plane's thrust reversers immediately, which is customarily done on slippery runways to help the plane stop, the pilot waited until fifteen seconds after touchdown to engage the thrust reversers. As a result, the jumbo jet barreled down the runway, crashed through a runway perimeter fence, then slid across an airport road, through an airport perimeter fence, and onto an adjacent public roadway, killing a six-year-old boy in a vehicle and injuring twenty-two people.

After digging into this flight's accident report, I shuddered over how much the circumstances reminded me of TACA Flight 390.

The NTSB concluded the probable cause of the accident was "the pilots' failure to use available reverse thrust in a timely manner to safely slow or stop the airplane after landing, which resulted in a runway overrun." The NTSB also named other causes that contributed to the crash, for which both the City of Chicago, as the airport owner, and Southwest Airlines were responsible. It was determined that Southwest policy failed to require landing distance assessments or to apply adequate safety margins. The City of Chicago Department of Aviation (DOA) was found responsible for operating a substandard runway. Specifically, runway 31C lacked an engineering materials arrestor system, also known as an arrestor bed, which is typically constructed of crushed gravel and is intended to stop an aircraft during a runway excursion. Had the arrestor bed been constructed five years earlier when it had been required by the FAA, Flight 1248 may never have crashed.

Some might wonder why the FAA hadn't enforced its authority

and ensured the City of Chicago complied with this requirement, but as someone who's worked in government for years, I knew exactly why. Government workers can't be effective unless political powers allow them to be. Investments in airport infrastructure can be expensive and disruptive to operations. In this case, there was no available space to construct an arrestor bed, which meant the city would have had to either purchase additional land or implement eminent domain procedures. Both options were contentious and controversial.

Six months after the Southwest 1248 accident, the FAA announced it would begin to require landing performance assessments based on real-time weather, wind, and runway conditions before planes could land on runways affected by rain or snow. In addition, the FAA called for a 15 percent safety margin to ensure sufficient landing distance at the time of arrival. Per this new FAA regulation, if the 15 percent margin of safety wasn't available, a pilot was not to land the plane there.

I was surprised to learn this type of requirement hadn't always been in place. In engineering, we often incorporate safety factors to protect against unforeseeable conditions, variability in test data, or the unpredictability of human behavior. But after the FAA made its announcement, it encountered considerable opposition to its proposal from the airlines. Clearly, the airlines believed the changes would disrupt operations, resulting in fewer on-time arrivals and more unhappy customers. Ultimately, the pushback was driven by the airlines protecting their bottom line. But how could the FAA, the industry's regulatory agency, the designated protectors of the public, let themselves be bullied? How could they ever justify delaying critical airport and policy changes, knowing those changes would save lives?

As I read more facts on the crash of Southwest 1248, I found the politics and bureaucracy involved staggering. And disheartening. On August 31, 2006, the FAA officially backed off of issuing the mandatory specifications. Rather, they published an interim Safety Alert for Operators (SAFO) 06012, which would allow airlines to voluntarily adopt these FAA recommendations. In other words, airlines could choose to adopt a policy that would require landing performance assessments. Or not. Airlines could choose to enforce safety margins on wet runways. Or not.

The entire purpose of accident investigations is to learn from an incident and make safety recommendations intended to prevent a recurrence. In my experience leading after-action investigations, I learned the value of identifying the gap between an intended outcome and what actually happened. These after-action assessments provide invaluable insight and lessons to be learned from real-life events. We know worst-case scenarios happen all the time, and if an incident occurred once, there's a high likelihood it will occur again. The intent for those of us who investigate these types of events is to build a more resilient system that will reduce the chances things will go wrong in the same way twice. The recommendations borne of these investigations should never be discretionary. If we're not going to learn from these accidents and make intentional change, why bother investigating?

Although the FAA governs only US aviation, when it comes to the safety of potentially millions of customers worldwide, why shouldn't all airlines follow recommendations that result from all plane crashes and other near-miss aircraft events? TACA Airlines opted not to voluntarily adopt the SAFO 06012 recommendations, and the government of Honduras ignored its responsibility to construct an arrestor bed for Toncontín's runway. Had either of them taken responsible action, TACA Flight 390 probably wouldn't have crashed.

I believe many of life's events are cyclical and that we're doomed to repeat tragic incidents unless we proactively take measures to prevent them. The lessons learned from Southwest 1248 offered a shining opportunity to improve aviation safety. But the decision-makers at TACA Airlines and Toncontin Airport once again placed profits over safety and chose to do nothing. As I read the Southwest 1248 accident investigation report, I was overcome by a deep sense of sadness and helplessness at the lost opportunity that could have resulted in saved lives. It was chilling to identify the moment when someone made a conscious decision that set the wheels in motion for the tragedy that killed my brother and four others.

PART III

Miracles Every Day

"The shell must break
before the bird can fly."
—Alfred Lord Tennyson

14

Breaking News

"Have you heard the news?" Mom blurted over the phone. "A plane landed on the Hudson River!" It was January 15, 2009, around 1 p.m. Mom had just returned to El Salvador after spending the Christmas holidays with Freddie and me in LA, and I was surprised to hear from her so soon.

I had taken the rest of the week off from work, and I was in the kitchen washing the lunch dishes. "On the Hudson River? In New York?! Are you sure you heard right?"

Six months after Cesare's death, Mom was still very fragile, and it was clear she had little patience for my questions. "Of course, I heard right," she snapped. "I'm watching it on CNN."

I hurried over to the TV in the den, and as the screen came alive, I gasped. There was a massive plane floating on the Hudson River. Right on the edge of New York City.

"Mom, let me call you back." I suddenly felt nauseated, and my fingers tingled. Like TACA Flight 390 just six months earlier, this aircraft looked unnaturally misaligned and grossly out of bounds. *Had it also overshot the runway?* I wondered.

All the TV stations were reporting the breaking news that US Airways Flight 1549 had landed in the Hudson River. I stared at the screen, watching the disabled flying machine floating in the frigid waters with over 150 people standing on its outstretched wings, huddled together waiting to be rescued. The scene was so odd, I thought it could have inspired a Salvador Dali painting.

News reporters said Captain Chesley Sullenberger, "Sully," appeared to have successfully landed on the water with no apparent damage to the plane and no casualties. "Miracle on the Hudson," they were calling it. It certainly looked like a miracle.

I sat fixated on the cockpit, which seemed to be undamaged. Its haunting, dark windows stared back at me. There was no need for equipment to free the pilots; they walked out of the plane on their own. This story wouldn't require body bags.

Except for some light snowfall in the morning, Flight 1549 had glided through crystal clear blue skies on a spectacular New York City afternoon. I wondered how different things might have been if Cesare had encountered similar conditions.

For the next few days, I was riveted by the miracle story, and I read every report I could find. US Airways Flight 1549, an Airbus A320-214, had departed LaGuardia International Airport on Runway 4 at about 3:25 p.m., en route to Charlotte's Douglas International Airport with a final destination of Seattle-Tacoma International Airport. I looked up the technical specifications for the aircraft and learned that, except for a different type of engine, Sully's plane was essentially the same plane Cesare flew.

About ninety seconds after departure, creating a scene that could have been straight out of a Hitchcock movie, the jumbo jet struck a flock of birds. Immediately, Flight 1549 lost both engines, and its ascent began to slow.

The whole truth would only be made available years later. Sully described how, faced with the worst crisis of his professional career, he reverted to the three basic rules of aircraft emergencies: Aviate. Navigate. Communicate.

Recognizing both engines were permanently damaged, he knew the only way to maintain control of the massive jet was to glide through the air. This may sound like a peaceful flight over New York City, but the plane was descending very quickly, not unlike a hotel elevator dropping at a rate of two floors per second.

Control tower cleared him to return to LaGuardia, from where the flight had taken off just three minutes prior, and New Jersey's nearby Teterboro Airport also cleared Sully for an emergency landing. But given how quickly the plane was descending, Sully knew he wasn't likely to make it to either facility. He thought fast and decided rather than risk trying to make it to either of those airports, he'd do something highly unconventional. He decided to land the metal bird in the Hudson River. Sully knew that if he was successful in landing on the surface of the water, the aircraft's partially full fuel tanks would keep the plane afloat long enough to allow the passengers time to evacuate. He also knew the ferry boats on the Hudson would be able to respond immediately, saving precious time in rescue efforts.

The accident investigation report for the US Airways 1549 accident states that during the short flight, Sully's only message over the intercom to the passengers was, "This is your Captain, brace for impact!" Then he raised the nose of the plane slightly as he began his landing. He knew this was a tricky maneuver given the speed of the plane and how fast he was approaching the water. The jumbo jet needed to land slightly nose up as it touched the surface of the water. If the nose was too high, the tail would hit the surface of the water hard and break apart. Too low, and the plane would descend into the water. For the aircraft to land smoothly and stay afloat, Sully had to keep the nose slightly above horizontal level, so he ignored the aircraft's automation urging him to *pull back* as the plane continued its rapid descent. The plane hit hard and sliced through the river, sending water splashing violently against the plane's nose and windshield. Then, slowly, Flight 1549 came to a floating stop. Sully had landed the massive jet with no casualties. What a glorious day it was, a day an aviation mishap didn't end in disaster.

Although I'd never met this Sully, I was bursting with pride in him. I spent that afternoon watching the scenes as the little ferry boats circled the crippled plane and rescued the passengers who'd been crowded into lifeboats and those still standing on the wings, knee-deep in cold water. I was uplifted by the fact that sometimes plane crashes can result in happy endings. And everybody loves a story that ends so happily; I could practically hear Hollywood knocking on Sully's door.

Sully is and always will be a hero—because things worked out as they did. One thing that worked in Sully's favor was that he had options, albeit few. He could have returned to LaGuardia or flown to Teterboro Airport as directed by air traffic control. But he decided to land on a river, an option no one in a control tower would have ever offered. The decision when and where to land is the pilot's call.

Ten years later, I watched Sully recount the worst day of his career on *Sully Unfiltered: Captain Sully's Minute-by-Minute Description of The Miracle On The Hudson*, featured online on *Inc. Magazine*. Poised and articulate, he revealed that after eliminating all other options and deciding on the Hudson River landing, he just knew he'd be successful. He had no doubt. Sully explained that in the moments before he landed the plane, he was entirely focused on what he needed to do each step of the maneuver, as pilots are trained to do. They're driven by adrenaline and a breadth of information provided by education and experience. And fueled by immeasurable confidence. Excess confidence.

In his 2009 memoir, *Highest Duty: My Search for What Really Matters*, Sully wrote: "I survived in part because I was a diligent pilot with good judgment, but also because circumstances were with me."

Skills and luck. Sully had both on that day.

The Flight 1549 crash landing triggered an immediate investigation, with the NTSB acting as the lead. Parties to the investigation were the Federal Aviation Administration, US Airways, US Airline Pilots Association, and Airbus Industrie, among others. These agency representatives were tasked with investigating the events that occurred in the 208 seconds that transpired from the moment the birds hit the plane until Sully landed the plane on the river. Despite Sully's accumulated 19,633 flight hours, he would be judged for his actions during those 208 seconds. For pilots, a legacy can be built or destroyed in a few final seconds.

As part of the investigation, flight simulations were conducted in an attempt to duplicate the conditions of the flight. The pilots performing the simulations had the benefit of hindsight and were fully briefed on the maneuvers prior to the simulation, so in a way, these exercises were a bit like taking a test after reading all the answers. Although twenty simulations were initially performed with varying runways at LaGuardia and Teterboro Airports, five of those tests were discarded due to flawed data. Eight of the remaining fifteen simulations resulted in crashes when the planes failed to reach the runways due to the rapid descent. These simulated flights had been conducted by seasoned pilots who already knew the circumstances surrounding Sully's landing, pilots using flight simulators that were safely grounded on land. Yet in barely over half of their attempts, the planes reached the LaGuardia or Teterboro runways safely.

The simulations proved it was possible to land at one of the runways, but they also proved that the odds were almost even that the plane would or wouldn't make it back to either LaGuardia or Teterboro and land successfully.

Sully heard about the experiment's 50 percent success rate, and, perhaps concerned that the NTSB might be considering pilot error, requested a reaction delay variable be included in the simulated exercise. Having survived the accident, Sully had the ability to advocate for himself. It was only because of his request the NTSB added a simulation test in which the test pilot would divert the plane after a thirty-five-second pause, which represented the approximate time it took Sully to weigh his options and make a decision on where to land. This additional test resulted in a simulated crash. By calling for an augmented test, Sully may have saved his reputation. Without the new results, the NTSB might have concluded pilot error.

Immediately following the incident and for years afterward, the media played a big role in elevating Sully to hero status. Day after day news outlets shared stories heralding the brave, competent pilot who seemed to have achieved the unachievable. Sully was elevated to an almost godlike status, and his humility only augmented his image as he credited his crew, the air traffic controller, the passengers, and the rescuers for the event's happy result. The story and the man were one package, and the public couldn't get enough.

Because of the swell of public support for Sully and his choices that day, the NTSB may not have been too keen on concluding pilot error, even if they'd been leaning that way. A much higher court of popular opinion had already concluded Sully had miraculously saved every soul on that plane. The outcome had been successful, whether through skill or luck, or a combination of both, and the NTSB had nothing to lose in supporting the conclusion that the US Airways 1549 crash wasn't pilot error. Years later, Tom Hanks played Sully in a 2016 film based on the true story of the river landing, which further elevated the heroic pilot's status. With Hollywood itself deeming Sully a hero in a film that showed him emerging victorious over an industry initially bent on making him the villain, Sully's legend status was locked.

Sully *was* a hero and clearly a skilled pilot. But had there been no miracle that day, had the Hudson River landing ended tragically, Sully's decision to land on water would have been judged harshly by many. A catastrophic outcome would have been deemed pilot error, and Sully's legacy would be that of the antihero.

Just one month after the Miracle on the Hudson, news of another plane crash flashed across television screens, a third aviation disaster in just over six months. I'd just arrived home from work and was in the kitchen nuking leftovers for dinner. Farfalle al pesto. My favorite. Freddie was still at the office, so it would be dinner for one.

I heard the news anchor on CNN. "We interrupt this program to bring you breaking news." I rushed over to the TV in the den just in time to hear that Continental Connection Flight 3407, operated by a regional air carrier Colgan Air, had fallen from the sky over a neighborhood near Buffalo, New York. The aerial view of the crash site showed the remains of the aircraft engulfed in flames, as the news anchor reported the Bombardier Q400 aircraft had crashed onto a home. The wineglass shook in my trembling hand.

Three plane crashes in eight months. The proverb "bad things happen in threes" came to mind, but I shook it off as a silly superstition. I'd been on at least a dozen flights since Cesare's crash, and I'd survived them all. I certainly didn't need to let superstition upset my already rattled mind. I had literally flown halfway across the world in that time. Should that be my new perspective? That a flight taken was a flight survived. Are we all survivors in a crash avoidance existence? The randomness of life and death.

I stood there as the news of the Colgan crash unfolded. All forty-nine onboard and one person in the house had been killed. Flight 3407 had been on its way from Newark's Liberty International Airport to Buffalo Niagara International Airport, and although cleared to land, the plane slammed into the private home five miles shy of the runway. The images on CNN showed the remains of the plane, its distinctive blue tailfin stamped with the white-and-gold globe. That night, fifty families would be notified that their lives would never be the same again. If they were lucky, that news would be followed by the deliveries of the bodies of those they loved and their loved ones' belongings.

Thinking about those families, I thought about the little red Teflon document bag I'd received just before Christmas. Cesare's personal documents had finally been recovered, and the American embassy in Honduras mailed them to me. Alone in my room, I'd slowly opened the Federal Express package to find inside a manila envelope containing an official US embassy–issued inventory list of items in the package. The list was folded over a small red Teflon document bag and secured with a large rubber band. Tears streamed down my face as I ran my trembling hands over the bag, hoping that holding something Cesare had touched on that tragic day might somehow connect me to him.

I reached to unsnap the buckle and noticed dried blood on the inner ridges of the snap in an area that would be difficult to clean even for a sensitive embassy worker trying to remove all signs of trauma. I sobbed. It had to be Cesare's blood.

Inside the bag were Cesare's personal identification documents, including both his US and El Salvadoran passports, his pilot's licenses, and several flying identification certifications. At the bottom of the stack of laminated IDs was a 2x3 card with the image of a bearded Jesus accompanied by a short prayer on the back, *Jesus, protect us in our travels.* Who had placed this card amongst Cesare's things? Mom? A girlfriend? Cesare?

No, not Cesare. He wasn't the religious type. Was he?

I decided that at that moment, it no longer made a difference. I stared with contempt at the little card in my hands. Whoever had slipped the card in the Teflon bag, apparently compelled by faith in a bearded man, would be sorely disappointed that it failed to protect its owner in his travels. Maybe Jesus had been on vacation that day.

This would be the last of Cesare's things we would recover. There would be nothing more.

That night, I wondered if any family members of the Colgan Flight 3407 victims would ever receive their version of little red Teflon bags.

A year after the Colgan crash, the NTSB's accident report concluded pilot error, specifically the pilot's inadequate response to the plane's stall warnings. As the plane began to enter a stall, the pilot pulled back on the control column, which pitched the nose up and increased the stall speed. The report confirmed the appropriate response would have been to lower the nose by pushing the control forward and applying full power. In the last two minutes of flight, the plane pitched and rolled from left to right and back again. In those final terrifying moments, as the plane plunged 800 feet, passengers endured excruciating forces of almost 2G, which would have resulted in the feeling of being pinned down by something double their weight. I felt nauseated as I thought about the terror passengers must have felt as the plane went down.

The investigation's report indicated the pilot's actions had exacerbated the crash. But the report also exposed several alarming safety issues that potentially set the stage for the tragedy. The report found the airline had demonstrated a lax training program, maintained inaccurate training data records, and implemented flawed checklist procedures. Colgan also failed to enforce federal regulations and policies created to prevent pilot fatigue and failed to maintain a sterile cockpit, which ensures crews abstain from casual conversation in order to focus on the task of flying the plane.

The investigation revealed that during training exercises, the captain had demonstrated substandard performance for

landings and takeoffs. Why, then, didn't the airline require reme-
dial training prior to allowing him to continue flying as the pilot
in command? Colgan had a reactive safety culture and chose to
ignore high-risk behavior. The problems were systemic.

Besides being the most recent deadly airline crash in the US,
the Colgan incident was significant in other ways. After Flight
3407, the FAA capped the hours a pilot can fly in a seven-day
period. They also increased the minimum flight experience hours
from 250 hours to 1500 for commercial licensure and enacted
stronger training record background checks for pilots.

It became clear to me there were several causes that led to
the Flight 3407 crash, and I didn't see why they shouldn't all have
been clearly named as being responsible for the disaster. Yet the
investigation laid blame on the pilot.

Just two weeks after the Colgan Flight 3407 crash, Sully testified about his accident in front of the US Congress Subcommittee
on Aviation, a subcommittee within the House Transportation
and Infrastructure Committee. At the hearing, Sully sat at a
table, flanked by the crew members who had been involved in the
harrowing flight emergency. When introduced by the ranking
chair of the committee, Sully received a standing ovation from
legislators and members of the public for his heroism. He knew
that with his hero status intact, he, probably more than anyone
in the moment, could be the most effective advocate for change
in the aviation industry, so he took the opportunity to call for
balance among accountability, safety measures, and training.
Sully reflected on a time long ago when there had been sufficient
investments in infrastructure, training, and recruitment and
retention of talent. Then he closed with a warning, "We must
develop and sustain an environment in every airline and aviation organization a culture that balances the competing needs of
accountability and learning. We must create and maintain the
trust that is the absolutely essential element of a successful and

sustainable safety reporting system to detect and correct deficiencies before they lead to an accident."

Because the next aviation accident wasn't likely to result in a miracle on a river or anywhere else.

15

Near Misses

For about a year after Cesare's crash, I suffered from post-traumatic stress disorder. I knew what I was experiencing couldn't be as awful as what a soldier might feel after living through combat, so I felt embarrassed using the term PTSD in reference to my circumstances. I hadn't even been anywhere near Cesare's plane when it crashed into the embankment. But I learned from articles published by the American Psychological Association, Mayo Clinic, and several other sources that dealing with a loved one's traumatic and sudden death can cause serious psychological disturbances. And I was certainly experiencing psychological disturbances.

I was often spooked by cars getting too close to mine on the freeway. I became jittery at the sound of an ambulance, fire truck, or police sirens. I was haunted by nightmares. I felt out of balance, vulnerable, unsettled. I lived in fear of everyday aspects of life that I'd previously taken no notice of, like driving in traffic and being in large crowds and of course, traveling by plane.

In April 2009, Freddie and I went to Palm Springs to relax for a weekend. It was a toasty 100 degrees in the desert, so the day after we arrived, we decided to spend time in some nicely air-conditioned shops. We had just left the hotel in our BMW and sat at a red light, engaged in deep conversation as I looked at something on my Blackberry. When the light turned green, we rolled into the intersection.

Boom! We never saw the car coming.

Both front airbags exploded before us, and the fabric bags completely blocked the windshield, which turned the inside of the car dark. My Blackberry flew out of my hand, hit an airbag, and projectile shot back at me. I turned away, but it smashed into the top of my head. Meanwhile, the collision had hurled our car into oncoming traffic, which sent us crashing into the back of a Toyota Camry and thrashing our bodies back against our seats. Our car spun, then finally came to a stop just shy of a light pole across the street.

Within seconds of the second crash, the inside of the BMW was flooded with smoke. That could only mean one thing—the car was on fire! I was suddenly overwhelmed by claustrophobia as I felt myself pinned to my seat.

Not only was I surging with panic for myself, in that moment I pictured Cesare strapped down in the crushed airplane cockpit. My mind became a jumble of past grief and current desperation. We had to get out. We had to get out right away or we were going to die.

"Are you okay?" I asked Freddie. He nodded, looking stunned. Then I yelled, "We have to get out!"

I pulled the door handle on my side of the car, but the door wouldn't budge. And now I was beginning to hyperventilate, which I knew was very bad because the car was full of smoke. Was my panicked breathing going to lead to a faster death?

The impact had activated BMW's Assist system, and a voice floated into the car. I couldn't understand the words, but I was sure I was hearing a voice from above. I mean, *really* above. Above, as in the universe. Or the heavens. A higher being. Was I already dying? Was I dead?

Despite the fact that things were happening as fast as a ball flying around a pinball machine, everything seemed to be happening in slow motion. I wondered if Cesare had experienced something similar. Just as Cesare had been, Freddie and I were confined to a tiny cabin, and I believed my husband and I were headed for the same horrible fate. But unlike Cesare, at least I wasn't going to die alone. I looked over at Freddie, who was coughing as he bashed at his door. I tried once again to push the door with all my might, and this time it creaked open. I squeezed out through the narrow opening of our crippled vessel and gulped a huge breath of fresh air as I let my eyes focus on the situation all around me. That's when I realized the car wasn't on fire. The smoke wasn't smoke at all but dust particles coming from the airbag.

Now that the car door was open, the chalky dust was beginning to dissipate, and the airbags hung limply from the panel. Three crumpled vehicles sat this way and that in the middle of the road. As I stood there dazed, Freddie jumped out of the car and ran to me. We hugged until we were interrupted by *the voice*. "We have detected a collision. Are you okay?"

I got back in the car on the driver's side and told the voice we were fine.

Good Samaritans stopped to see if we were okay. "Hey," said a bearded man in a baseball cap, "I witnessed it all. That man in the silver Mercedes ran the light."

The ambulance came screaming to the scene, and paramedics checked Freddie and me for injuries. Despite the bump on my head from the flying Blackberry, I was okay, so the paramedics said I was free to go. Freddie was shaken but also determined to be unhurt. The tow truck hauled our broken car away, and Freddie and I went back to the hotel to hold each other, sprawled across the plush king-size bed. My head gently rocked with every rise and fall of his chest.

"You know, it could have been so much worse," I said.

"It was pretty bad."

"I know, but we did walk away relatively unscathed. Goes to show that life can change on a dime. Today was another reminder."

We were full of what-ifs. What if we'd been T-boned or our

vehicle had flipped over? What if we'd plowed into a vehicle with a higher center of gravity, like an SUV? What if our car had actually caught on fire and we'd been unable to get out? There are countless scenarios that could have resulted in a catastrophe. These what-ifs provided fodder for my PTSD-induced frazzled nerves.

In May 2009, I was once again on a TACA red-eye flight from Los Angeles to San Salvador, on my way to Cesare's one-year memorial to be held in a couple of weeks. Freddie was bogged down at work; he'd just been promoted to Assistant Director at Public Works. As the number two executive in the agency, he was juggling his calendar and he'd join me the following week. As a family member of a TACA employee, I was traveling on a complimentary round-trip ticket. Shortly after the crash, TACA committed to granting this company benefit for life to Mom, Mars, and me. But this kind of ticket didn't guarantee a seat, and with the passage of time ticket sales representatives, who pro-cessed the complimentary tickets, were soon forgetting why these tickets were being granted to us, which meant securing a seat was getting harder. So I no longer felt like a VIP on TACA Airlines. And ever since my meeting with Mojica, I'd felt the company had quite simply betrayed us. The only saving graces were the employees, pilots, and flight attendants who still treated us like family.

The flight was oversold, so I was seated in a rear-facing jump seat at the front of the plane adjacent to the first-class restroom. Jump seats are meant for flight attendants who must be seated and strapped-in during takeoff, landing, and turbulence. They're not meant for passenger use. In fact, the seat folds out and back again, there are no pull-out tables, there's no button to recline the seat back, and there's certainly no plush padding. But I recognized the airline was bending the rules by letting me sit in a jump seat at all, so I was grateful. I was also anxious.

Since the crash, I'd had a keen awareness of which part of the plane I was in whenever I flew. My little jump seat faced the back of the plane, so I was essentially flying backward, which gave me an uneasy feeling, a kind of lack of control. I found my discomfort ironic, since we, as passengers, essentially relinquish all control when flying in an aircraft, no matter where we're seated.

As I sat uncomfortably with a clear view of the first-class cabin, my mind wandered to my research of the extensive damage Flight 390 suffered in this area of the plane. Upon impact, the nose landing gear had penetrated the belly of the aircraft. The avionics bay, a plane's electronic equipment, which is located below the cabin floor, was thrust upward into the area of the nose landing gear. As the cabin floor was punctured from below, the first four rows pushed the first-class restroom and galley forward, crushing the cockpit from behind. The electrical box was found about seventy feet from its original location near the main passenger exit door at the forward end of the cabin, the very same one located next to me. And the overhead compartment bins collapsed onto the seats from the front of the plane back to row eleven.

I stood to stretch my legs, hoping it would also stop my brain from torturing me, when the captain stepped out of the cockpit and approached me. "I'm told you're Cesare's sister," he said, shaking my hand. "My name is Johnny Torres, pleased to meet you."

"My pleasure, Captain."

"Please, it's Johnny. How are you? Are we treating you well? Sorry about the jump seat."

"Yes, thank you, all is great. Please don't apologize. I'm happy to be on the flight."

"How's the family?"

"We're doing okay. Thank you for asking."

"I can't believe it's almost a year," he said, shaking his head slowly, small pools of water collecting in the corners of his eyes. "And your mom? How's she doing?"

"As well as can be expected." I shrugged. "She's struggling."

"I just can't imagine. I have two daughters . . ." His thought trailed off as he looked over my shoulder toward the exit door behind me. "We all loved Cesare. He is truly missed. It never should have happened." We stood in the quiet galley with only a dark blue curtain separating us from the passengers.

"Can I ask you a question?" I said. "You don't have to answer."

"Shoot."

"Have you ever landed in Toncontin?"

"Yes, many times."

"So you have the additional required training and certification."

"Yes."

"Is landing there as bad as everyone says it is?"

Johhny sighed. "Listen, takeoffs and landings are the most dangerous parts of any flight. Any flight. But Toncontin is in a completely different league. There's a whole series of maneuvers we have to make quickly," he said, snapping his fingers several times. "The workload is pretty intense, all compressed within a few minutes." He shook his head. "It's just such a complex airport. It leaves no room for mistakes. That landing must be perfect. Each and every time. And you know, none of us are perfect 100 percent of the time. You're either perfect or you're lucky. Hopefully, you're both."

I nodded, recalling Sully saying that his miracle on the Hudson had been both skill and luck.

Johnny tapped his head. "You know, when I'm coming in for that landing at Toncontin, there are a million things running though my head. Calculations, a heightened awareness for my surroundings, traffic control, sights, sounds. There's a pit in my stomach, and my fingers tingle. It's an explosion of adrenaline." He shook his head. "Each and every time." There was an awkward silence as he reflected on something, his eyes shifting side to side. Then he placed his hand gently on my shoulder and said, "You should know that Cesare was a great pilot. He was one of our best. If he couldn't nail the landing that day, none of us could have."

None of us could have.

I nodded and said, "Thank you."

He squeezed my shoulder and said, "I should get back."

"Yes, you have a plane to fly," I said with a grateful smile.

He bent down to kiss me on both cheeks. "It really is great to meet you."

I was genuinely surprised by Johnny's candor. I was, after all, a complete stranger. But he'd been wonderfully familiar and relaxed with me and had allowed me into his world, the final seconds of every nerve-racking landing at Toncontin Airport. The sensations, the sounds, the sights. Vivid descriptions that brought me along as he relived the experience and added a human dimension to something I'd only experienced analytically through technical words in reports. I couldn't help but wonder if he and other pilots who knew Cesare ever contemplated, "It could have been me that day."

The plane dipped, and I sat down quickly. My instinct was to grab the armrests, but I realized my little jump seat didn't have any. Not having anything to grab onto during this little bout with turbulence shot my nerves again. A voice crackled over the speakers, "The captain has turned on the seat belt sign."

I buckled my shoulder harness, making sure it was taut, then I gripped the straps. As I looked out the little window trying to ground myself at 35,000 feet, I realized there was a familiarity to Johnny's story. Years earlier, Cesare had told me that on a flight where he was first officer, the plane was on approach to Toncontin, and they'd been cleared to land. Cesare said, "There was this thick fog, and we couldn't see the runway. We were on descent, and the captain looked desperately for the runway lights. But there was a thick, soupy mix of water droplets in the air, looked like a dense cloud. We couldn't see a thing."

When telling the story, Cesare seemed to be reliving the experience. He looked over my shoulder into a far-off space. "I could see the captain's eyes scanning the white cloud outside the windows as he strained to see the damn runway lights. I was looking too. But you know, he's the captain, and he's gotta see them himself. He's gotta be confident that he's good to land. And then, 'Captain, there they are!'" He'd pointed at an imaginary runway. "Man, I could feel the pit in my stomach and blood pumping in my neck. But as much as the captain tried, he still couldn't see the lights, so we did a go-around. The captain did the right thing. You don't land the plane unless you're absolutely sure you can stick the landing. There are just too many lives at stake."

I thought about what that meant—that during a landing, all those lives are in one person's hands. As a passenger on a flight, my responsibility is to buckle my seat belt and steer clear of the aisles. Beyond that, I'm expected to keep myself occupied—take a nap, read a book, watch a movie. Despite the fact that we had a pilot in the family, in all the flights I'd taken before Cesare's crash, I'd never given much thought to what was going on in the cockpit. I was probably like most other passengers who took for granted that the pilots were smart, trained, healthy, rested, and sober, and that they'd get us to our destination safely. I assumed every airline's pilots had received proper training, that all airlines implemented adequate safety policies and developed schedules that allowed for appropriate rest periods. I assumed aircraft and airport infrastructure were properly designed and maintained and regulators provided the appropriate oversight to this complex interconnected system.

I never thought about near misses or close calls. I never questioned any of it. But today, I recognize that near misses and close calls happen every day, and I've simply been a consumer living in the dark.

After the crash, our family received emails from a handful of TACA pilots sharing concerns about what had happened to TACA 390. Pilots wrote that they'd been critical about the condition of the runway. Crewmen had complained about hydroplaning after construction work performed at the airport and said these grievances had been brought to TACA's attention, but that TACA did nothing. There had been other close calls at Toncontin, including borderline landing overruns. TACA was aware of all of them but took no action to prevent more near misses and disasters.

One of Cesare's colleagues, a former TACA pilot, said, "We've all been in a car when we experience a close call and someone yells at us, 'Watch out!' And we miraculously maneuver the vehicle and avert a collision. That's a near miss. No big deal. All's well that ends well, and we go on our merry way. I think about that stormy day of the crash. I think about that day a lot. About the days that led to that day. About the conditions that everyone was well aware of. I think about Cesare, a great pilot, a wonderful captain, and even better friend. He was about to have an accident, and no one was there to yell out, 'Watch out, don't land!'"

No one was there, not on land nor in-flight, to tell Cesare, "Watch out!" I now think about this every single day.

16

One Year Later

It was May 28, and that afternoon, I had picked Freddie up from the airport. He had juggled his work calendar to arrive in El Salvador in time for our wedding anniversary, just days before Cesare's memorial. I had arrived the week before, and we were now feasting on take-out lasagna and Caesar salad from one of Mom's favorite Italian restaurants. A bottle of pinot noir stood guard in the center of the candlelit table at Mom's house where she and Mars joined us for this feast extraordinaire.

"Happy anniversary, sweetie," I said, as I leaned over to kiss Freddie. Four years earlier, I said "I do" to my beloved.

Mom said, "I'm sorry you had to celebrate like this."

"Mom, what are you talking about? What more could we ask for? A wonderful celebration with the people I most love. I can't think of anywhere we'd rather be."

I sat and stared at my husband as he chewed a bite of sourdough bread. He was wearing a long-sleeved Ralph Lauren oxford shirt that had belonged to Cesare—a sign that after a year, Mom was finding ways to move on. She'd allowed Mars and me to pack up Cesare's condo and contact a real estate agent, and she'd recently sold his Mini Cooper and put his six-month-old Porsche Cayman up for sale. "It's practically new," she told the dealer. "It has less than a thousand miles on it." Mom had also started sorting through Cesare's clothes and though she found it painful, she forced herself to give some of them away. "I'd like Freddie to have these," she'd whispered one day, pointing to a box she'd filled with some of Cesare's shirts. Freddie proudly accepted the generous gift with a steady flow of tears.

The year had been hard on all of us, and Freddie had been especially patient as I'd traveled back and forth handling Cesare's affairs and immersing myself in my obsessive search for the truth. I smiled at my husband, so supportive, so understanding. So handsome in the candlelight. In good times and in bad.

On May 30, 2009, my family arrived at the cemetery chapel for Cesare's one-year memorial, where Mom had arranged for a 5 p.m. mass in his name. The turnout was hearty but considerably smaller than the crowd that had showed up for his funeral. The sensationalism of the story had passed, and now only our closest friends and relatives attended. At the altar, the man dressed in robes read a passage from a book. He then pointed to the congregation who parroted back a response. I felt no connection to this religious tradition, so I sat alone in the back of the chapel away from everyone. Freddie, ever the sweetheart, sat next to Mom at the front of the church.

My family spent a lot of time designing the perfect mementos to give to the memorial's attendees. The night before, we'd spent hours in a simulated production line assembling the keepsakes: little crystal angel lapel pins clasped in light blue ribbon and attached to a card with a picture of Cesare in Paris, the Eiffel Tower as a backdrop. The card read:

> *On May 30, 2008, Cesare wore his flight wings*
> *and a little crystal angel pin*
> *on his pilot uniform lapel.*
> *This little crystal angel is a*
> *symbol of guidance and protection.*
> *On this day, we would like to share with you*

the angel that accompanied Cesare home
and although we are left heartbroken,
we are also left with a flicker of hope and faith.
Today, Cesare is our angel.

The message was inspired by French philosopher Pierre Teilhard de Chardin, who said we're all spiritual beings having a human experience. In other words, we're all immortal on a never-ending journey, and once our time here on earth is over, we'll return to our spiritual home. I'm not sure I believe that, although I will admit it does have a feel-good vibe. But as I watched people pinning their angels onto their clothing, I couldn't help remembering that when we were kids, whenever someone broached a supernatural concept that seemed a little out-of-this-world, Cesare would hum the theme song to the *Twilight Zone*. I could almost hear him humming it as I sat there pondering the ever-after.

I sat in the brightly lit chapel and thought about how amused Cesare would be at all the things I'd tolerated in the year following his death, church for one. But he'd really get a kick out of my obsession with learning all I could about aviation. For decades, Cesare had tried to engage me in his passion for flying, to no avail. Yet in this past year I'd immersed myself in the painful process of trying to interpret aviation jargon, technical data, analyses of mechanical components, and endless hours of complex research. Even in death, he got the last laugh.

The mass finally ended, and the churchgoers exited in my direction. As the crowd moved toward the back of the church, I was overcome by the sight of a flight of little crystal angel pins that seemed to be floating on the lapels of friends and family. Each person went home with their own personal guardian angel, the faint sunlight reflecting off each winged crystal pin, the light blue ribbon standing out as a sign of hope against the dark hues of their clothes of mourning.

As the crowd of attendees disbursed, Mom, Mars, Freddie, and I walked the short distance over the gravel walkway to the cemetery. It was dark now, almost 7 p.m., and the full moon lit up the night sky. I had with me a ten-pack of sky lanterns I'd purchased on Amazon. I'd seen countless pictures of lanterns being released into the sky, sometimes hundreds of them, magical celebrations of floating candles, and I thought it would be very cool to light up the skies of San Salvador with our own version. What I'd failed to consider was that none of us gathered had ever used one of these flying lamps.

No big deal. Freddie and I were engineers—*How hard could it be?* I wondered as I ripped open the pack and spilled its contents onto the walkway next to Ceare's brass plaque. Freddie pulled the instructions out while I bent down to gather the lanterns and other pieces that were strewn about.

"Uh, the instructions are in Chinese," he said, squinting in the dark. "But it looks like, yeah, there are step-by-step illustrations. Okay, unfold the lantern and open it."

I took one of the lanterns and aired it out like a trash bag. One of the seams split open, and I could see the lantern was now useless.

"Shoot, I tore it. Pass me another one," I said. I tried to be more careful with the next one and took more time to gently expand it. The paper was surprisingly lightweight and delicate. Some of the most beautiful things are the most fragile, I thought. And so easily broken. Standing here in Cesare's final resting place, I couldn't help thinking of the fragility of life.

Freddie handed me what appeared to be a wax cube with a hole in it. "It looks like this is the fuel cell, and you're supposed to loop it onto the wire at the opening of the lantern." I followed his instructions carefully and wiped the excess wax off my fingers. Then he said, "Is it secure? Okay, then, now you light it." He passed me the box of matches we'd brought with us. Then he held the lantern sideways, and I lit the match and held it to the wax cube, feeling the heat of the flame as it inched toward my fingers. I had to use several matches before the fuel cell finally ignited. The flame burned the tip of my fingers, and I yanked my hand away. Then the lantern shifted position, and the flame quickly caught the paper on fire. The lantern dropped to the ground now fully engulfed in the blaze, and Freddie furiously stomped on it.

"You guys look like Keystone Kops," my sister said, giggling.

And suddenly, my mother burst out laughing, a hoarse guffaw echoed across the green gardens of the cemetery. It was the most beautiful sound I'd heard in over a year. And then each of us added to the chorus of cough and giggles. Laughing tears streamed down our faces, and I was filled with hope that maybe our healing had begun. And I hoped that just maybe, Cesare, on some supernatural plane, was sharing the moment with us.

On our third try, we lit the wax cube safely and precisely, allowing the hot air to fill the paper lamp, and we set it free. We stood there holding each other as we watched the little paper beacon take flight and shine in the night sky. Its glow was as brilliant and radiant as any of the night's twinkling stars.

17

Air France 447

The next day, we awakened to the devastating news that Air France 447, an Airbus 330-200, had taken off from Rio de Janeiro headed for Paris and vanished three and a half hours later. I sat glued to CNN as the news unfolded. Planes didn't just disappear. Something terrible had happened. I watched the heartbreaking images of the dumbstruck relatives of the 216 passengers and twelve crew members rushing to Charles de Gaulle Airport for an initial briefing, their faces distorted with pain and disbelief as they crumpled to the ground. The flight information screen at the arrivals terminal still eerily displayed the flight as *RETARDE*. Delayed.

The images of the family members rocked me because I knew I was witnessing the dawn of a new hell for these people. Despite the thousands of miles that separated us and despite that I didn't know any of them, I could feel their suffering. As our family was just taking a turn toward healing, I was watching their awful journeys just beginning. There was no protecting them or soothing their pain. There was no hitting fast-forward to skip over the agonizing parts of the timeline. There was no magic wand that could make any of it go away.

Five days later, Freddie and I were packing to return to Los Angeles, but Freddie was doing more news watching than packing. "Hey, they seem to have found some remains," he said.

For the first five days, the news reports hadn't revealed much more than, "nothing new to tell," and I'd thought that not knowing must have been utterly unbearable. But now, I wasn't so sure. Now they'd found remains. Was confirmation that someone you love had indeed died in an airplane disaster even worse?

The first images on CNN were of floating airplane parts—a galley kitchen and a passenger seat. Then a briefcase. I sat on the edge of the bed and watched as the camera panned to the largest piece of evidence in the water: the tailfin of the doomed aircraft, clearly showing Air France's signature red and blue stripes. Whatever glimmer of hope anyone might still have had was extinguished right then. Although it was still unclear what had happened specifically, clearly the aircraft had suffered a cat-astrophic end.

I sat stunned in front of the TV as CNN reported on the recovery of body parts pulled from the choppy waters. It was now more than a year after our tragedy, but watching this story unfold caused me to relive the anguish. As I watched the breaking news, I realized I was forever a member of a group I didn't want to belong to, a group whose membership was founded on pain. And our heartache would be revived every time another plane crashed and added new members to the group.

Once back in Los Angeles, I continued to follow the story of the mysterious plane disaster. For months there would be no new information, no black box, and my interest began to wane. But I wondered if there were any people like me among those family members, people obsessed with finding the truth of what hap-pened that day with whatever little information they had.

Then, almost two years after the plane went down, the black box was recovered from the ocean floor, and the investigation went ahead full speed. In 2012, the investigation was finally con-cluded after revealing the aircraft stalled due to poor weather, equipment malfunction, and pilot error.

The accident report exposed problems with the plane's pitot tubes, which are located in the front of the fuselage and measure the plane's speed. Both Air France and Airbus were well aware of the pitot tube problems prior to the crash. There had been at least sixteen related incidents recorded as far back as 2005, with nine of those incidents occurring in the twelve months prior to the crash. The two companies involved had plenty of time to work to fix the equipment malfunctions, but just as TACA had, decision-makers in the company chose to ignore the alarming data.

During the accident investigation, even the design of the Airbus cockpit had come under scrutiny. Airbus used a single-handed sidestick aircraft control similar to an arcade joystick. By contrast, Boeing uses yoke controls, U-shaped handles on a control column. In a Boeing aircraft, the yokes for the captain and the copilot are mechanically linked so each pilot is aware when the other is pulling or pushing on the yoke controls. As a result, there's no opportunity for mixed signals. But Airbus's sidesticks aren't linked and operate individually. At one point during the Air France 447 event, both pilots tried to operate their sidesticks at the same time, which inadvertently canceled each other's actions and accelerated the stall.

Years ago, when TACA Airlines transitioned its fleet from Boeing to Airbus, Cesare spent time in Toulouse, France, home to Airbus headquarters, as he trained to fly the new aircraft. He quickly fell in love with the new plane, enamored by Airbus technology and automation. He loved the glass cockpit, which I learned was a sexy way to refer to an aircraft cockpit that boasted electronic flight controls with LCD screens and 3D navigation systems. Upon his return from Toulouse, he told me that with a slight shift of the joystick, the technology could interpret his commands and adjust the plane's moving parts. Cesare said it was as if the machine was an extension of a man, and they operated as one. For him, there was an elegance and grace about flying an Airbus plane, something he hadn't experienced before. I found his selection of words almost poetic. Cesare's love affair with flying deepened with every technological advancement.

Typically, the conclusion of an accident investigation is the end of the story. While a conclusion often leaves at least some interested parties unhappy with the findings, the ending of an investigation often does provide some semblance of closure. But in this case, a group of 384 family members of the victims of Air France 447 banded together to demand justice for the deadliest crash in the history of Air France. The families weren't satisfied that the blame should lay solely on the pilots with no responsibility accepted by either Airbus or Air France, so for over a decade, they navigated a legal gauntlet of trying to hold the two French corporations accountable for their actions, or lack thereof.

In October 2022, thirteen years after the tragedy, the families got their day in court. The French criminal court opened a corporate case of involuntary manslaughter against both Airbus and Air France. This was historical. No other aviation company had ever been accused of killing its passengers. The experience would test not only the families of the victims but the alliance between the two corporate giants.

Despite having to relive the agony of losing their loved ones, family members remained strong and united, faithfully attending the legal proceedings to ensure the victims remained at the heart of the case. But Airbus and Air France turned on each other. Airbus blamed pilot error, while Air France argued the aircraft had presented inaccurate data when the pitot tubes malfunctioned.

I rooted for the families and admired their resilience. But after a nine-week trial, they were cheated out of their chance for justice. In April 2023, fourteen years after Air France 447 plunged into the sea, the judge declared the evidence fell short of placing blame on either corporation. The judge *did* conclude there had been several acts of negligence by both Airbus and Air France in failing to take action regarding the faulty pitot tubes, but deemed that a case hadn't been made to fault that negligence for the crash.

I couldn't help but resent the disparity between the bar used to judge corporations and the one used for individuals. I was stunned to see that despite so much evidence pointing in multiple directions, the industry norm remained *blame the pilot.* I'd followed the trial on and off for fourteen years, and when some of the loved ones of the victims stepped up to microphones to make public statements my heart broke.

Danièle Lamy, who lost her son in the crash and heads an association for families of victims, said, "For the powerful, impunity reigns. Centuries pass, and nothing changes. The families of victims are mortified and in total disarray."

Philippe Linguet lost his brother and is now vice president of the victims' group. He said, "A day of infamy, mourning, sadness, and shame."

Brazilian Nelson Faria Marinho lost his son, an engineer who'd been on his way to Angola to work on an oil exploration job. Marinho said, "With all the accidents, all the tragedies, the first thing they do is blame the pilot, which isn't true. It is a killer plane, and they didn't correct it."

The verdict represented the end of a long, drawn-out battle, but for the families of the victims of Air France 447, hearing "not guilty" was like losing their loved ones all over again. The fact that justice wasn't served only made the experience that much more heartbreaking.

In the end, the disaster led to several technical and training changes. As with so many aviation reforms, however, people had to die before negligent parties sat up straight and took notice. It all came too late for those who boarded Air France 447 trusting that responsible parties were guarding their safety.

At least in the Air France 447 case, two aviation behemoths were forced to enter a courtroom to face their victims and defend their actions. And they were publicly admonished for their negligence. But as always, the spin doctors are always standing by to start the cleanup of corporate reputations.

After the Air France 447 case ended, I wondered if I was witnessing the birth of a wave of change. Maybe that trial would be the catalyst for families coming together with the courage to hold large corporations accountable for their actions. Maybe we'd finally be able to hold religious institutions, pharmaceutical companies, financial organizations, gun manufacturers, and the aviation industry accountable for their disregard of life driven by corporate greed. Change takes time, but maybe the miracle of Air France 447 was that it set off the flutter of a butterfly's wings.

18

The Report

A little over a year after the crash, Mars received a WhatsApp message from a friend, a high-level executive at TACA Airlines. It read: *You should know that the accident report will not be made public. But your family has a right to know what the investigation has concluded. I can get a copy of the report for you.*

Mars offered to meet him for coffee, but he replied, *No, I think it's best if I get it to you via messenger service.*

No meeting in person. A messenger would deliver confidential documents never meant for our eyes. It felt as if we were in a spy game in which we'd soon be privy to top-secret information. Why wasn't the Salvadoran Civil Aviation Authority going to make the report public? What was there to hide? Maybe something was being covered up. Maybe this report was going to be the smoking gun I'd been hunting for months.

That afternoon, a messenger delivered a CD to my sister's home. I loaded it onto my laptop as nausea crept from my stomach up my esophagus. Mars didn't feel she was ready to handle reading the report, so I offered to dissect the information and let her know what I learned. I opened the 119-page Word document and realized that, just like the preliminary report, this document was also undated. I found that so strange. Then I suspected the omission was because the report was never meant to see the light of day. But how was this withholding possible? This report was the result of an official accident investigation led by the Salvadoran Civil

Aviation Authority. The investigation team had been comprised of members of several international organizations, including the FAA and the NTSB. Certainly, someone involved or someone on the periphery would eventually ask, "When is that report due to be published?"

Clearly, TACA Airlines had a copy. Who else had access to the document? And who decided not to make the accident report public? I wanted to know how anyone could get away with concealing the truth behind this kind of investigation. But I already knew the answer to that question. TACA Airlines was a powerful company able to leverage its might. And I was convinced that's exactly what it did, used its commanding influence over the Salvadoran Civil Aviation Authority, an agency of El Salvador's government, to prevent the report from being published. It's the only thing that made sense.

The report confirmed what the TACA executives had told Mars and me months earlier: Their research had found the cause of the crash to be pilot error, that the probable cause of the crash was Cesare's decision to land the plane. Now that I saw their claim in front of me in writing, I was enveloped by a heavy sadness.

But the report also named other causes that contributed to the crash, including a lack of a runway emergency safety area and reduced braking efficiency due to lack of pavement grooves, which resulted in inadequate runway drainage. *I'd been right.* The runway infrastructure was substandard. I wondered if this was why officials wouldn't release the report. It would be damning to the owner of Toncontin Airport, the Honduran government.

The report also stated the pilot didn't calculate the actual landing distance prior to commencing the approach. But as I dug deeper, I discovered this conclusion was misleading. The report referenced the FAA's Safety Alert for Operators (SAFO) 06012, which was issued as a result of the Southwest 1248 runway excursion that occurred in 2005 at the Chicago Midway International

Airport. But the report confirmed that TACA never adopted the SAFO recommendations; neither the requirement to assess the actual landing distance nor the additional 15 percent safety margin were ever implemented by the airline. In fact, the report stated that neither the Quick Reference Handbook nor any operations bulletins available to the flight crew during in-flight operations had flagged the need for a calculated distance. If the policy was never in place and pilots weren't aware of the need to assess an actual landing distance plus a 15 percent safety margin, it would be completely inappropriate to lay on the crew the responsibility for this nonexistent protocol. Had TACA done the right thing and implemented the safety recommendation provided by the FAA, the crew would have performed the calculation and added a 15 percent margin to the landing distance required. After running the numbers based on real-time weather conditions, Cesare would have realized he needed almost 6,000 feet to land the aircraft. 6,000 feet. The available landing distance of Toncontin's Runway 02 was only 5,410 feet. At that point, it would have been obvious to Cesare and anyone else at the helm that he should not land the plane. Cesare would have had the information to make a sound decision. Only then.

Based on the FAA recommendations and under the best of conditions, the damn runway was too short! This was the smoking gun!

I stared at the screen. There was no denying what I was seeing. This was why the report would never be released. TACA outright ignored the FAA's recommendation and chose to gamble with people's lives.

The investigation identified flaws in the runway as a cause of the crash, but TACA chose to pin the accident on Cesare's so-called excess confidence, which they claimed influenced his decision to land. So they not only misrepresented and maligned my brother to cover their asses, they chose to hide the truth by never making the report public.

I was having trouble breathing as the massive weight of the truth set in. A simple calculation based on an FAA recommendation as a result of a real-life incident could have prevented TACA Flight 390 from overshooting the runway. The words and numbers in the damning report on my computer screen became a blur as I wept for my brother who was failed by the very organization he served.

Protect your little brother. Always protect your little brother.

The heaviness weighed on me like a boulder, so I set out on a long walk around the neighborhood park that was surrounded by beautiful lavender-purple jacaranda trees in full bloom. The same park where Cesare and I had tossed a basketball and swung on the swings, gripping the chains as we pumped our legs and reached for the clouds. Soaring, weightless.

"I won't let you fall," I had promised long ago.

But today, the park was quiet, and the swings hung idle. Still my promise remained steadfast. I headed back to Mom's house where my laptop awaited me.

Back at Mom's dining room table, I scrolled to the beginning of the report and read it from start to finish. The report included low-resolution transcripts with accompanying pictures of the flight path. There were a significant number of photos of the wreckage showing details of the aircraft components, and looking at them now I saw extensive damage to the fuselage I hadn't seen before. There were also photos of the inside of the cockpit, which had been crushed. The impact had caused the nose of the plane to crumple inward, but the passenger cabin kept moving forward, crushing the cockpit from both ends. It was hard to imagine this compartment holding two grown men.

Seeing the pictures now put an end to the comfort I'd given myself every time I'd pictured my brother after the crash. In my wild imaginings, the inside of the cockpit was intact. I'd pictured Cesare slumped over on the left chair in surroundings that were essentially unscathed, maybe just a few untethered belongings strewn about—manuals, eyeglasses, handheld devices, headsets. But these pictures showed the destruction of the equipment, mangled and misaligned, with exposed wires and broken parts tangled in an unruly web. Instrument panel parts had shattered and crushed into other components of the flight controls. The front panel had buckled and crumpled so the electronic aircraft screen, similar to the display in the dashboard of a car, now faced upward. The cockpit seats and sidesticks area were also described as having rotated ninety degrees clockwise, which would have forced Cesare and the copilot into a face-down position. This most likely explained Cesare's skull fracture; as the aircraft came to a violent stop crushing the cockpit, the pilot's seats, still in forward motion, rotated on their hinges, and Cesare's head slammed into the control panel. The images made clear my brother had experienced a violent end.

An airport security camera captured the flight's landing, and a still shot clearly showed water from the runway spraying the plane's wheels. The time stamp was 9:49:20. A snapshot of Cesare's final documented moment alive.

The view from one aerial photo of the wreckage showed the plane lodged in a ravine at an angle counterclockwise from the runway. Had the jet continued in a straight direction, it would have collided with the surrounding businesses and homes. It appeared as if the plane had been steered away from the structures atop the embankment. The picture showed the runway to the right of the broken fuselage with skid marks starting at about 600 feet from the end of the runway and continuing to where the asphalt met a ten-foot stretch of grassy area that led to the hinge point of the descending slope. There was no runway emergency safety area. There was no arrestor bed.

The caption for the skid-marked runway photo read: *A wire fence, supported by metal poles inset in a cement base, protecting the cliff edge, was broken by the passage of the airplane.* I couldn't help wondering why the cliff edge needed protecting but the jumbo jet racing down the runway did not.

I pored over the report, reading and rereading the raw data, the calculations, and dense technical jargon. I had to stop often to google the definitions of several aviation terms. With each discovery of some new revealing fact, I'd go back to the beginning of the report to read it again from the start, and with each excruciating pass, I relived the crash.

At Cesare's wake, TACA CEO, Roberto Kriete, had told me Cesare had shut the engines off right before impact. He claimed that, given the amount of fuel the plane was carrying, had it not been for Cesare's actions, the plane could have exploded. I scoured the report to find evidence of this claim but found nothing, so now I was weighing yet another person's words I couldn't be sure about. Barreling down the runway trying to stop the jumbo jet, I'm not sure Cesare would have had time to shut down the engines. I wondered if Kriete was speculating on that day, saying things he hoped would comfort me. I chose to believe he meant well, but he didn't know I'm a trained engineer with an obsessive need to get to the truth of a situation and that I accept nothing at face value. I was going to stay involved until I could verify every piece of information in the report.

Was it true the jumbo jet could have exploded? Through a Google search, I learned there are several factors that can result in an aircraft explosion due to a crash: speed, rate of descent, and the type of damage to the plane. But Cesare's crash conditions weren't conducive to an explosion. TACA Flight 390 had already landed on level ground, and its impact speed was about sixty-five miles per hour. Also, the wings, which is where the fuel is stored, never sheared off. Had the wings been ripped off the fuselage, the fuel vapor may have ignited, but given the speed of the aircraft and the location where it came to rest, an explosion was unlikely. I concluded Cesare never shut down the engines, but he saved a lot of lives by steering the plane away from the embankment. While I appreciated Kriete's attempt to ease my mind, I'm never comforted by words that aren't backed by evidence. The facts indicated my brother never cut the engines, but that he did act fast enough to save countless people who will never know he's the guy to thank.

The report described Toncontin Airport as a complex airport, the most complex in all of TACA's system, which required special procedures and training for day-to-day operations due to terrain, approach and departure challenges, weather conditions, and high-density traffic. Cesare had gotten all the training.

I was alarmed to learn there was no specific training required for first officers flying to Toncontin. This meant if the captain should become incapacitated, the plane would be in control of a first officer who would be gravely undertrained for the challenges posed by Toncontin Airport's unusual approach conditions. But why was I alarmed? I'd already seen evidence TACA had no problem putting its passengers and crew at great risk.

The accident report indicated that in 1998, Airbus set certain limitations on the use of their jets at Toncontin. For starters, they deemed runway 02 should not be utilized or if necessary, should be used only minimally. *What?* Yet TACA Flight 390 was dispatched from El Salvador with runway 02 as the planned runway for landing. At that time, Airbus also stated there should be no landing on a contaminated runway, *contaminated* defined as having more than 3 mm of water—or less than one-eighth of an inch, the thickness of two stacked pennies—on its surface. Given the weather conditions caused by Tropical Storm Alma and the lack of runway grooves, it's highly likely the runway met the definition of contaminated on that day. But the runway remained open, and the control tower cleared Flight 390 for landing.

As part of Airbus's 1998 stipulated recommendations, no landing should have been attempted at Toncontin Airport if tailwind was greater than five knots. Airbus also provided a physical landmark, the first perpendicular taxiway located 820 feet beyond the displaced threshold, as the appropriate touchdown point under wet conditions. Both the tailwind and appropriate touchdown point limits were exceeded that day. Cesare landed with a tailwind of ten knots and touched the plane down 490 feet beyond the designated taxiway.

In the voice recordings with the control tower, Cesare acknowledged the tailwind was greater than five knots, but he still decided to proceed with the landing. Why? Perhaps he'd landed at Toncontin previously under similar conditions, with a tailwind greater than five knots and hadn't had any problems. Pushing the boundaries of risk was part of TACA's culture. The message was: when in doubt, go for it.

For reasons I'll never be able to verify, Cesare, knowing the surrounding conditions that day, chose to land the plane. He expected to stick the landing. Cesare expected this to be like any other workday from which he'd return home safely. That one decision, founded on thousands of decisions by many others, would be what Cesare would be judged on.

As was the case with Sully's accident, the TACA 390 report described how the investigative team assigned to Cesare's crash analyzed whether any different actions by the crew could have resulted in a different outcome. But unlike during Sully's post-crash investigation, Cesare's possible alternative actions weren't played out on a simulator. Instead, the fictional scenarios were computed empirically. In other words, people behind a desk tinkered with data and developed theoretical analyses that made sense to them with the full benefit of hindsight. They conducted a theoretical exercise to counter a real-life event. Of course, in the world of make-believe, anything is possible.

The post-crash analysis included three hypothetical scenarios to determine whether there could have been an alternate outcome to TACA 390. In other words, was it possible TACA 390 could have stopped before overshooting the runway? The first scenario showed the plane overshooting the runway just like TACA 390 had, and it resulted in a crash. The other two schemes averted a collision with a short length of runway remaining. These hypothetical successful outcomes imagined several changes to the actions of the pilot: earlier application of brakes, earlier nosewheel touchdown, and maximum braking action until full stop. What struck me was the randomness of the exercise. When dealing with multiple variables such as a plane landing, there are countless iterations that could be run. Thousands even. I couldn't help but wonder who decided to stop at three scenarios, and who decides which three those should be? Were those three chosen because they all pointed to pilot error? Why not run additional hypothetical scenarios with, say, an appropriately designed runway safety area or an arrestor bed? Because it wouldn't support the conclusion of pilot error but might call attention to alarming systemic problems?

While the report concluded pilot error, it was also peppered with recommended corrective actions that pointed to members of the aviation system other than the pilot. TACA had failed to provide a safe environment for its employees, which the report confirmed by stating that post-crash, the company should implement policy requiring pilots to perform landing distance assessments with a 15 percent safety margin prior to all landings, and that they should provide training for dispatchers and pilots requiring landing distance assessments that incorporate a minimum 15 percent safety margin. In other words, the policies borne of the Southwest 1248 crash would finally be implemented. The report also recommended TACA perform a risk assessment of the complex airports in its system. I looked at the report and sighed. Would TACA do the right thing now?

One final recommendation pointed to regulatory agencies and certifying authorities, including the FAA and the large aircraft manufacturers, like Airbus and Boeing. The Salvadoran Civil Aviation Authority made clear aircraft designers should include in all operational documents used during preflight and in-flight procedures a clear warning of the need to apply a safety margin to landing distance calculations. I was stunned to learn the manufacturers had never included this type of warning in any of its documentation. Ever.

The recommendations in the report were heavily focused on the airline's lack of safety policies and lack of appropriate staff training, the shortcomings that led to my brother's crash. The report made clear the regulatory agencies and the aircraft manufacturers were also at fault. Still, TACA blamed the pilot. Then TACA hid the truth in a thick report they tried to make sure no one would see.

I supplemented the information in the accident investigation report with my own research, and I came across a quote by James Reason, a former professor of psychology, and renowned expert in aviation and human error. In his book *Human Error*, Reason writes something that counters TACA's conclusion of pilot error: "Rather than being the main instigators of an accident, operators (pilots) tend to be the inheritors of system defects created by poor designs, incorrect installation, faulty maintenance, and bad management decisions. Their part is usually that of adding the final garnish to a lethal brew whose ingredients have already been long in the cooking." A lethal brew indeed.

Reason isn't the only related expert who believed that flaws in an organization's defenses can allow accidents to occur. In her book, *There Are No Accidents*, Jessie Singer, journalist and expert in safe systems, concludes, "Pilot error is almost always a consequence of some flaw in the built environment. Even so, blaming human error remains the industry norm today." Reason and Singer offered me comfort.

If we conclude human error is the cause of a tragedy, we must include *every* related decision made by a human. In the search for the truth behind a tragedy like a plane crash, we have to open our eyes and examine the actions of every person responsible for organizational and bureaucratic compromises to safety. The report in front of me was riddled with such compromises, blatant screwups that led to my brother's death. A death that was completely predictable and preventable.

I closed the report and rubbed my eyes with the heels of my hands. I needed a break before I could carefully process what I thought I'd seen while taking in so much jolting information at once. A Post-it note peeked from a page of the report. I'd marked the page I thought held the final piece of the puzzle. After dinner, I'd open to that marked page and see if my suspicions had been right, that I'd marked the page that held the smoking gun.

19

The Smoking Gun

I've dedicated my life to protecting and safeguarding the health, safety, and welfare of the public by ensuring safe infrastructures. I've helped communities recover and rebuild following natural disasters, crafted design policies to address the effects of a changing climate on the built environment, and I've championed the tightening of building code requirements. As a member of the ASCE, I've advocated at the federal level for increased infrastructure investments to improve public safety. I've collaborated on the development of infrastructure report cards that assign letter grades, just like those used in schools. Aviation infrastructure in the 500-plus airports serving commercial flights in the US— terminals, runways and taxiways, air traffic control towers and equipment—received a grade of D+. Just shy of failure. And I've advocated for aviation safety and the timely approval of the FAA Reauthorization Act to secure adequate funding for aviation facilities. So it was a cruel irony that my brother died because of an unsafe structure, an airport runway.

The accident investigation report turned out to be a treasure trove of damning information. But in the midst of all the data and figures and finger-pointing, there was a section with information I realized was what I'd been desperate to find all along. Buried deep in a nondescript section of the report and titled *Aerodrome Operator*, I discovered a lengthy discussion of the conditions of the airport.

Call me, I texted Freddie, then paced the patio until my phone rang. I answered on the first ring and blurted out, "I think I found it! The smoking gun!"

"Hold on, slow down. What happened?"

"The runway was deficient!"

"Okay, break it down for me from the beginning."

I took a deep breath to slow my racing heartbeat. "The State of Honduras owns the airport but contracts with Inter Airports, who's responsible for operations and maintenance of the airport, including the runways. On December 12, 2003, Inter Airports was given a four-month deadline to fix the runway by placing an asphalt overlay on the existing runway and taxiway."

Freddie said, "Four months—that's doable. It's what our folks do to extend the life of a paved road so long as the base of the pavement is still in good shape."

"Right, but four years later the runway still wasn't fixed! How could anyone with any authority let that deadline pass without taking action? You know, like maybe shutting the airport for non-compliance? Something. Anything."

"So who was in charge of making sure the work got done?"

"Apparently, Public Works. On July 3, 2007, Inter Airports requested a deadline extension from the Honduran Director of Public Works. Inter Airports claimed structural deficiencies were found in at least 260,000 square feet along the length and width of the runway. Sweetie, I did the math. This meant thirty percent of the 900,000 square feet of the runway needed to be reconstructed before it could be considered safe. Thirty percent!"

"Wait a minute, this is all in the accident report? Are you kidding me?"

"Yeah. It's all there. It's buried deep in the report."

"Thirty percent of the runway was structurally deficient, and no one did anything about it? That would have shown up as potholes and buckling and uplifting. No one noticed any of that? Come on!"

"I know! And no, no one did anything about it for years. Where was the ICAO in all of this?

"Unbelievable. Did they get their extension?"

"They did. The Honduran government granted Inter Airports' extension request for 120 working days to complete the work, which meant for the next four months, planes would be landing on a runway that had been deemed dangerous four years earlier. It's even unclear whether Inter Airports' plan for the proposed improvements was ever reviewed and approved by any regulatory agency. If the plan was reviewed by someone, would it have been by someone technically qualified to review the design and construction proposal? The report is silent on all of that."

I heard a siren in the distance.

Freddie said, "And runway design's more than just slapping some asphalt over dirt. There's a whole slew of things to consider: the types of planes that'll land there, the weight, wheel configuration, all sorts of things."

"But wait, it gets worse."

"That's hard to believe."

"The work on Toncontín's runway went forward, and when it was finished, the General Directorate of Civil Aeronautics inspected the runway and found, and I'm quoting here, 'grave technical discrepancies in the work that posed a threat to air navigation.' That's straight out of the accident report! The newly improved runway was deemed a threat to safety. Can you believe this? That should have set off alarms. If I'd been in charge, the airport would have been shut down until the work was done right. But flights continued to be allowed in and out."

"This is incompetence at best. Negligence at worst," Freddie said. I could almost see him shaking his head.

"I'm not even done. So the GDCA sends a letter to Inter Airports asking for an assessment of the runway's actual surface friction coefficient, which makes sense, given that it measures an aircraft's ability to brake properly. Pretty important for short runways, right? This is an ICAO requirement for all rebuilt runways. But no one seemed to know what the actual coefficient of friction at Toncontin was! Apparently, the GDCA had been asking for the friction coefficient for five years!"

"Hold that thought. The pizza delivery guy's at the door. Be right back," Freddie said.

I hadn't been able to *stop* thinking. Working in public service my entire career, I knew what it was like to face resistance when requiring compliance for development and construction projects. Things often took longer than we wanted, so there was seemingly endless back-and-forth exchanges to get to a high-quality product, ensuring no corners were cut. But the GDCA's lack of oversight was blatant negligence by inaction. Perhaps the inaction was due to the GDCA's having no leverage to enforce the stipulated requirements. It was the only explanation I could think of for such a lackadaisical attitude on behalf of the agency overseeing safety conditions. Inter Airports was given until May 30, 2008, to comply with this request. May 30, 2008. The day of the TACA flight 390 tragedy.

"Okay, I'm back!"

"You're having pizza without me?"

"Yes, but it won't be the same without you," Freddie said. "Go on, you were telling me about this shit show."

"So the report also says that GDCA received complaints from TACA crews about Toncontin's runway conditions, post-repairs—complaints that the central axis was misaligned and that the braking action was deficient."

"What does that mean, the braking action was deficient?" he said with his mouth full.

"I don't know. It's not in the report. But I'm gonna go out on a limb and say that the TACA pilots were experiencing hydroplaning."

"Makes sense, we've heard about those complaints. The complaints should have alerted someone to take some level of action, not just take notes for some report."

"Yeah, the point is that TACA knew all along. Its pilots raised these concerns, and TACA's Flight Safety Office emailed the GDCA. The email says, 'It is worth stressing that the runway friction coefficient is very important due to flight crew reports that the runway's actual braking conditions are not very positive and in the wake of the coming rainy season we could be affected by a poor runway braking coefficient.' I'm assuming the Flight Safety Office reports to the Flight Safety Director. Mojica. TACA knew all along. And they too chose to continue to land on the flawed runway."

"This rabbit hole just gets deeper."

"Finally, on May 16, 2008, two weeks before the stipulated deadline, Inter Airports responded to the GDCA. Not only did they not comply, they said their contract didn't specifically require testing to determine the coefficient of friction. They claimed the Honduran government was responsible for that kind of testing. They argued the contract didn't specifically delegate these duties to the airport operator. In fact, they turned around and said the GDCA was responsible, something about the GDCA being the legal rep for the Honduran government."

"I'm not sure what to say. This is a complete clusterfuck."

"I know."

"I'm so sorry I'm not there to help you unravel this mess."

"You're on the other end of the line, and that's good enough for now. Love ya. Talk tomorrow?"

"Love ya right back."

Sitting on Mom's patio, it was hard for me to process all the damning information. How could I? Far-fetched and hard to believe, a cover-up on so many levels. It was something straight out of a Hollywood conspiracy movie. Landing a commercial plane on a rain-soaked deficient runway was an accident waiting to happen. Clearly. I was truly shocked. The incompetence I was unearthing was beyond anything I'd seen in my decades in public service ensuring compliance with safety regulations. I was amazed that this ineptitude, deliberate finger-pointing, and shirking of responsibilities didn't result in a greater number of deaths throughout the years.

The investigation report recognized these egregious short-comings and included several recommendations that fell within the wheelhouse of the Honduran government. The main recommendation directed the Honduran state to pass a law allowing it to delegate duties associated with airport operations and maintenance in order to comply with ICAO standards. It was absolutely staggering to think how the absence of this law had shielded Inter Airports for years from complying with safety requirements. I shuddered at the idea of how many other unsafe conditions existed at Toncontin because of this void in the legal system.

I'd now snapped the final piece of the puzzle in place and could see the entire picture of what happened the day my brother died. TACA neglected to implement appropriate safety policies and procedures, and the Honduran government blatantly failed to maintain its runway. That's why my brother's plane crashed. The tragedy happened because of intentional safety compromises made by many people for financial and political rewards. Reckless inaction. Immoral decisions. Ethical dishonor. A deadly years-long game of Russian roulette. Then one day, the loaded chamber aligned with the barrel.

I finally had the truth.

Vindication provided little consolation. For more than a year, I'd been obsessed with finding the truth about what really caused the TACA 390 tragedy. I'd immersed myself in facts, figures, technical research, and accident reports. The quest to find out the story behind the headlines had become my North Star. But now that I had the truth, I felt hollow. I'd been right all along about the real cause of the crash, but Cesare was still dead, and because of a cover-up, everyone in his orbit, his colleagues, people who studied plane crashes—hell, anyone who googled TACA 390—believed he was responsible.

I felt defeated. The truth meant nothing if it couldn't be heralded loudly and broadly. Only when I could show the world the truth would I exonerate Cesare. Only then.

20

Boeing's Fall from Grace

Nine years later, Boeing, another mega corporation, followed the TACA Airlines playbook. Conspiracy. Cover-ups. Corporate greed. Innocent victims. Heartbreak. Intrigue. Political intervention. The story had the makings of a blockbuster suspense thriller. Except it was real.

On October 29, 2018, Lion Air Flight 610 took off from Soekarno-Hatta International Airport in Jakarta heading to Depati Amir Airport in Indonesia. It was a routine domestic passenger flight that would take about an hour. On these brief flights, no sooner does the plane reach its necessary altitude than it begins its descent. But Lion Air 610 never reached its planned altitude. Thirteen minutes after takeoff, the plane crashed into the Java Sea, killing all 189 people on board. Besides a short blurb in the evening news, the story got little attention in the United States. But it certainly got my attention. Was I alone in wondering why a plane crash that killed 189 people wasn't getting major news time? Was I also alone in wondering how safe we truly were flying anywhere?

Less than six months later, on March 10, 2019, Ethiopian Airlines Flight 302 met a similar fate. This flight departed Ethiopia's Addis Ababa Bole International Airport en route to Jomo Kenyatta International Airport in Nairobi, Kenya. Flight 302 lasted six minutes in the air before plummeting to the ground. All 157 passengers perished. Two routine flights in just a few months, and they both crashed. Coincidence? I doubted it. Almost 350 people dead. Same aircraft model, both Boeing 737 MAX 8 aircraft, a new and "more efficient" 737 model the FAA had approved in 2017. This time, the news seized the world's attention.

Within days, despite no confirmed information of what happened, media speculation was rampant. Just like with Cesare's crash, everyone pointed to pilot error as the cause of both devastating crashes. Even Boeing executives publicly questioned the pilots' performance. Although I expected this kind of speculation, I was stunned Boeing would take such a public position. I was pissed off. And I was alarmed; why did they have the right to assign blame so early in an investigation? Right away, Boeing leaned on both airlines' questionable safety records surrounding pilot training and history of accidents. Boeing had a lot to lose if the spotlight turned on the design and manufacturing of the aircraft, so the company didn't waver from its position that the plane was safe; instead, they directed attention to the airlines, their policies, and of course their pilots, dead pilots who couldn't offer any conflicting information. Blaming the airlines was a perfect PR strategy; after all, in 2007, Lion Air, an Indonesian low-cost airline, had been banned by the European Union from flying over European airspace and wasn't allowed to resume business in the area until 2016 when it improved its safety rating. Ethiopian Airlines, the flag carrier for Ethiopia, also had a questionable safety record; since 1965, the airline had been responsible for over sixty accidents that claimed 494 lives. Both airlines' history worked well to support Boeing's storytelling.

Even though two crashes within five months on a newly developed aircraft seemed highly unusual, the FAA hemmed and hawed and refused to ground the fleet of 737 MAX planes as they continued to investigate, which meant while Boeing's people looked into the deadly matter, 737s continued to fly.

I had just finished reading investigative journalist Peter Robison's exposé on Boeing, titled, *Flying Blind: The 737 MAX Tragedy and the Fall of Boeing.* The riveting book explained that Boeing placed profits over safety and explored how the FAA had enabled Boeing to cut corners, which included years of the agency's prioritizing the growth of aviation business interests over safety regulation. It appeared the FAA's goals included removing obstacles, streamlining processes, and putting planes in the sky. But the FAA's mission is supposed to be safety, so where did safety fall in their hierarchy of priorities? According to Robison, in an effort to streamline the aircraft manufacturing process, the FAA delegated its independent oversight and aircraft certification process to the US aircraft manufacturer itself: Boeing. This allowed Boeing to dictate its own safety decisions. This would be like grading your own exam.

I couldn't put *Flying Blind* down as I learned the misdirection from the truth was a result of Boeing's army of full-time lobbyists exerting its mighty influence to try to pass laws to empower the manufacturer. *Corporate might.* These new laws created the close relationship between the FAA and Boeing that ultimately blurred the lines of responsibility and oversight, eroded clearly defined roles, and created a conflict of interest that resulted in self-certification of Boeing products. As far as I was concerned, the FAA had abdicated its duties, and Boeing was only too happy to assume the role of self-regulator.

It was China that took the lead and unilaterally grounded its fleet of 737 MAX planes. Before long, other countries did the same. It was only when compelled by international pressure that the US grounded the flawed aircraft.

Just days after the crash of Ethiopian Airlines 302, Representative Peter DeFazio, the head of the Committee on Transportation and Infrastructure, opened an investigation in search of answers. I watched the July 2019 hearing on C-SPAN and rooted for the victims' families as they testified before Congress, demanding answers. Their message had reached the highest echelons of US government, and I was happy to learn of this because Representative DeFazio had been in Congress for many years and had extensive knowledge of the airline industry. I was hopeful the families and the rest of the flying public would finally learn the truth.

It was March 2022, and I was headed to Sacramento for my first in-person business meeting of the California Board for Professional Engineers, Land Surveyors, and Geologists since the pandemic. I'd been appointed by California Governor Gavin Newsom in August of 2020, and as a board member I was responsible for protecting the public by ensuring our licensees provided professional services in a competent and ethical manner. This included the enforcement of laws to protect against contract violations, fraud, and negligence within the professions our board represented. For almost eighteen months we, like so many other organizations, had worked virtually, and because my appointment occurred during the COVID lockdown, I had met my colleagues only via a computer screen, so I was excited to finally connect with these people in person.

I had snagged a seat in the exit row on a Southwest Airlines flight and was just settling in. After securing my seat belt, I pulled the laminated trifold card from the seat pocket and realized I was on a Boeing 737 MAX 8 plane. I felt a shiver.

After grounding the fleet of flawed aircraft for twenty months, the FAA lifted the ban at the end of 2020, after approving significant changes to the flawed aircraft made by Boeing. I'd read that Southwest resumed flying the MAX 8 in March 2021. The US carriers had been motivated to get the planes back in the sky as soon as possible.

I snapped a picture of the trifold on my iPhone and texted it to Freddie. *Hey, I'm on a MAX 8. But they claim they fixed them.*

My phone immediately buzzed with a reply. *They lie.* I knew he was trying to ease my nerves with humor, but I agreed. Boeing had lied. DeFazio's investigation, Peter Robison's book, and several news outlets all reported that Boeing had designed a faulty plane and tried to cover it up. They concealed the truth from the FAA, the pilots, and their customer airlines. Then they knowingly allowed the majority of their defective MAX 8 aircraft to take to the skies. It sounded awfully similar to the lies and cover-up I discovered following the TACA 390 crash.

I found Boeing's behavior egregious. And terribly disappointing. This was *Boeing*, a once proud and respected American company. Boeing manufactured Air Force One. We trusted the life of the President of the United States to this company. They were one of our own. Made in America.

But this new version of Boeing revealed executives had been reckless and negligent and had displayed a complete disregard for human life.

The flight attendant addressed those of us in the exit rows. "As a passenger in this section, you have a crucial role to play in ensuring the safety of all passengers. In the event of an emergency, you are responsible for opening the exit door, clearing the area, and assisting fellow passengers in exiting the aircraft quickly and safely. I need a verbal yes that you understand and accept this responsibility."

As each of us sitting in the coveted exit row parroted our concurrence, my hearty *yes!* stood out for its enthusiasm. Another flight attendant stood in the aisle demonstrating the emergency procedures with exaggerated facial expressions and hand gestures that made the passengers laugh. Although I had to admit, these were clever ways to engage passengers with an exercise that would otherwise be ignored, I found it counterintuitive to send an important safety message using comedy. It seemed to undermine the seriousness of a potential life-and-death circumstance.

Our plane was now barreling down the runway, and I looked out the window. It had been over a year since I'd been on a plane, and I tried to settle my pulse by focusing on the buildings whizzing by. As we became airborne and began to level out, I thought of all those people who had been on Lion Air 610 and Ethiopian Air 302. At this point in their flights, they'd had only minutes left before their aircraft plummeted to the ground. I envisioned the sheer terror they must have felt being jostled on the plane, not knowing that a faulty software system had seized control of their aircraft, and their pilots struggled to override the artificial intelligence. Man versus enormous machine. Six agonizing minutes as it oscillated and pitched through the air at full throttle. I wondered if this catastrophic thinking about plane crashes while flying on a plane was my way of somehow staying in control. As I crossed the aircraft threshold and relinquished all control to the flight crew, was my brain trying to counter this loss of command by showing no fear of thoughts?

I tried to distract myself by opening my laptop and reviewing the handful of disciplinary cases our fifteen-person board would be discussing the next day. Our board would be required to enforce appropriate disciplinary action to any of our licensees who had violated the professional laws and regulations. Discipline could range from probation to revocation of a professional license to jail time. I thought about Dennis Muilenburg, the CEO of Boeing, and an aeronautical engineer himself. Where was the accountability for his actions? I recalled watching the hearings on TV in October 2019, when Muilenburg was summoned to testify before the Committee of Transportation and Infrastructure. Dressed in a blue suit and tie, Muilenburg walked into the crowded hearing room with only a quick glance toward where the families of the victims sat. Greeted by a cacophony of camera clicks, he took his seat at the table facing the panel of legislators on the dais. Even faced with mounting evidence of wrongdoing, he still defiantly denied having knowledge of internal documentation that revealed Boeing's concealment of safety concerns. From the beginning of the hearings until the closing words by Representative DeFazio, there was not a word of accountability from Muilenburg. There was never so much as an apology to the families. As an engineer myself, I was taught that my ethical responsibility was to protect the health, safety, and welfare of the public I served. As Muilenburg addressed the committee and denied any knowledge of his staff's actions, I thought about how grossly he'd failed that mission.

At the time, I wept for the families of the victims of Lion Air 610 and Ethiopian Air 302 as they rose and confronted Muilenburg, displaying poster-size pictures of their loved ones. I wept for their pilots, too, who Muilenburg blamed for the crashes, despite knowing Boeing had produced a flawed aircraft. At the committee hearings, Muilenburg was now forced to look at images of those whose lives he had cut short.

Boeing's 737 MAX debacle was best summed up by Representative DeFazio in his closing remarks, "Pressures from Wall Street have a way of influencing the decisions of the best companies in the worst way, endangering the public, jeopardizing the good work of countless, countless hardworking employees."

There it was. Corporate greed. It all boiled down to the greedy dollar.

This debacle resulted in millions in financial losses for Boeing due to non-operation. As a result of conspiracy charges brought by the Department of Justice, Boeing was forced to pay $2.5 billion in fines and compensation to the airlines and the victims' family members. And Muilenburg was fired. But much more significant than that, lives were ended and hearts were broken. All because Boeing made a bad decision to withhold the truth and keep the planes rolling off the production line without correcting a known flaw.

The flight attendant announced the start of our descent, and I closed my laptop and stored it in my briefcase. It wasn't long before we touched down in Sacramento. I grabbed my phone and texted Freddie.

Landed safe and sound. Announcing a safe arrival had become a common practice between us after Cesare died. But Boeing's egregious actions had made us even more aware that a safe flight is never guaranteed.

Standing in line on the jetway at the American Airlines terminal at LAX, I typed a quick text to Freddie. *Boarding now. Will text when we land. Love ya.* My phone immediately buzzed back with a heart emoji. It was January 24, 2024, and I was headed to Miami to attend an ASCE board meeting. I was flying on an Airbus 320, which I knew because I'd checked the website when I booked the trip. Just two weeks before, Alaska Airlines Flight 1282 had suffered a sudden decompression six minutes after taking off from Oregon's Portland International Airport en route to California's Ontario International Airport. The evening news showed frightening images of the passengers still in their seats buffeted by strong winds, oxygen masks dangling from the ceiling, the night sky clearly visible through a hole in the fuselage. The gaping hole appeared to be the shape and size of a door. A door that appeared to have blown out. Or had there been an explosion? Did someone open the door mid-flight?

As the news dribbled out, I learned that the aircraft was a Boeing 737 MAX 9, the next generation of the tainted MAX 8. It was reported a door plug had blown out. So now the traveling public would be educated on the specifics of door plugs, which are panels installed to replace emergency exit door openings. Now TV stations showered the viewing public with information and endless pictures of door plugs. But I wanted to know how this latest safety failure could have happened, especially on the heels of the 737 MAX 8 crisis. One would think the massive financial damage and erosion of trust after the MAX 8 crashes would have compelled Boeing to clean up its shop. One would think. But they didn't. Corporate behemoths like Boeing and TACA still determined to play high-risk games in the name of profit.

For now, I felt safer not flying Boeing. But I wasn't an anomaly. NBC News reported that passengers were intentionally avoiding flights scheduled with 737 MAX aircraft. Some travelers were avoiding Boeing planes altogether.

After 9/11, we travelers adjusted to standing in long TSA lines and removing our shoes and belts in the interest of heightened security. Now post–737 MAX crashes, were we also going to have to check for aircraft types and locations of door plugs in order to feel safer about flying? Could we ever trust the airlines again? Could we trust the FAA to fulfill its mission as safety regulator, or would it continue to live up to its reputation as a "tombstone agency" for historically failing to address safety concerns until there was a loss of lives?

Boeing's fall from grace cost them plenty. Financial impacts came in many forms: profit losses as a result of the MAX 9 grounding, compensation to other airlines, lawsuits brought by passengers, and a dramatic drop in Boeing market value. We've probably yet to see the end of the fallout. The 737 MAX 8 crashes and the MAX 9 door plug failures are perfect examples of cascading events that align themselves in what safety experts call the "Swiss cheese" failure mode, which refers to when holes in organizational defenses allow for catastrophic breakdown. Just as with TACA Flight 390, these failures didn't just happen. The breakdown in organizational defenses began decades earlier when Boeing made a conscious decision to "keep those planes rolling off the assembly line" even at the expense of safety. The disasters weren't simply pilot errors. We, as traveling consumers, should all understand that most often people don't fail, systems fail.

21

Playing with Fire

In January 2020, the 2019 novel coronavirus, later officially named COVID-19, was declared a public emergency by the World Health Organization. I first heard about the virus on the evening news, but I was only half listening. This was news coming from clear across the world, and I didn't think it had any bearing on my life. I had much more personal burdens on my mind.

Only five months earlier, my mother had died of a ruptured aortic aneurysm, a broken heart. I think part of her died when, shortly after the crash, two TACA pilots rapped on her door with the devastating news, "We're so sorry. Accident. Crash. Cesare. Dead." When my mother's legs buckled from the weight of the grief, the two men held her up. The wailing. The anguish decimated her spirit and body for the next eleven years. By 2019, she was completely worn out, and when the end came, it was swift.

After returning to California from El Salvador, I was emotionally exhausted from wrapping up Mom's affairs. Closing out a life. There was so much to do, everything from preparing our family home for sale to closing financial accounts. At the end of each day, my mother's beloved pups, Cookie and Mimosa, greeted Mars and me with sloppy kisses and tails wagging. But then they returned to stand guard at the front door waiting for Mom to come home. They too were mourning. Mom was the center of our orbit, and we all felt lost without her.

I simply didn't have any energy to think about anything more than what was going on in my family; I certainly wasn't going to concern myself with a virus all the way over in China. But viruses don't respect geographical borders. In just a few weeks, it hopped from Wuhan to Northern Italy to the shores of New York, and it quickly became clear to me that the world was nearing a disaster of epic proportions.

In March 2020, the WHO declared COVID-19 a pandemic, and the world went into lockdown. Enormous components of societies literally shut down. The COVID disaster was like no other I'd ever experienced. As the Deputy Director over Public Works Emergency Management, being able to manage this crisis from my operations center at home was a blessing. Following Cesare's death, my home became my safe haven, gave me a sense of peace, and sheltered me from the virus.

Ever since Cesare's death, if there was a logical reason to view a scenario through an aviation-related lens, I found myself doing so. And I knew that COVID would be the catalyst for long-lasting effects on the safety of air travel. What I couldn't imagine at COVID's early stages was the magnitude of these effects.

The US government quickly imposed a travel ban in an attempt to avert community spread and uncontrolled entry points for the pathogen from China and Europe, which meant airlines literally stopped flying. So sure enough, in 2020, the airline industry experienced a 60 percent drop in passenger travel. Airlines were financially hemorrhaging, and I knew it was a matter of time before furloughs were implemented. I'd later learn that many of Cesare's former colleagues from TACA Airlines had been furloughed; some had even been stranded in foreign countries on layovers when the travel ban was implemented. So I was happy to see that the US government put up over $74 billion to bail out the US airlines. This was an attempt to save the nearly eleven million jobs supported by the aviation industry and the $1.8 trillion aviation contributed to the US economy. But while the aviation industry didn't lay off staff, they did offer financial incentives for early retirement that essentially reduced staffing levels, the very thing the $74 billion bailout was meant to prevent.

The pandemic also further stressed the FAA, which was already experiencing challenges. A revolving door in its top leadership position. An erosion of trust in its reputation. The need to upgrade outdated technology. And perhaps the most critical issue, air traffic controller shortages. These staffing shortages represented more than just flight delays and passenger inconveniences. In order to meet demand, the few air traffic workers available were often forced to work erratic schedules and ungodly overtime hours, which added stress to an already extremely stressful job. Without a proper support system, air traffic controllers admitted resorting to shortcuts and risk-taking.

Pilot and air traffic controller shortages resulted in an alarming uptick in near misses in 2023. In fact, a *New York Times* investigation showed that in the month of July 2023 alone there were forty-six alarmingly close calls, both in-air and on runways, that could have resulted in deadly outcomes. Forty-six! The NTSB concluded these were a result of human error, both from pilot and air traffic controllers. There were jumbo jets racing toward each other on the same pathway along crisscrossing runways. Aircraft cleared to land on the same runway on which another airplane was just departing. Planes in the sky mistakenly directed into the airspace of another. So many examples . . .

January 13. Kennedy International Airport. An American Airlines jet crossed the wrong runway, placing it directly in the deadly path of a Delta Airlines plane taking off as it barreled down the runway, forcing the Delta jet to abort its takeoff.

February 4. Austin International Airport. A FedEx cargo plane was cleared to land on the same runway where a Southwest Airlines flight was cleared for takeoff. The two aircraft came alarmingly close to a catastrophic collision.

July 11. San Francisco International Airport. An American Airlines aircraft raced down the runway at 160 miles per hour on takeoff, narrowly missing a Frontier Airlines jet that had mistakenly edged into its path.

These are just a few frightening examples of planes with passengers inadvertently placed on deadly collision courses, and all those examples were labeled human errors by the NTSB. Human errors by air traffic controllers, pilots, or both. All frontline workers. But what's responsible for so much human error? I'd argue we're still witnessing an erosion of safety-related standards in the aviation industry, a causal chain of events triggered by a deadly virus years before.

This increase in near misses should serve as warning signs sounding blaring and thunderous alarm bells. These events were so disturbing they made headline news in 2023.

New York Times: "Airline close calls happen more often than previously known."

Slate Magazine: "Everyone seems to agree a major plane crash is coming."

Politico: "We were very lucky: Near collisions spark new worries for air travel."

New York Times: "Delta 1943, Cancel Takeoff': Wrong Turn Results in Near Miss at JFK."

Business Insider: "Dangerously Close Near Miss Accidents Across the US Are a Wake-up Call to the Aviation Industry."

The year 2024 wasn't any better for the aviation industry. With increasing frequency, news outlets released alarming stories that exposed aviation safety gaps, stories of airplane parts falling from the sky, runway excursions, aircraft collisions on the tarmac, engine malfunctions, and severe turbulence cases resulting in serious injuries, all involving major US carriers. The news of near misses, technical mishaps, and faulty aircraft manufacturing rang alarm bells that called attention to the need for more rigorous oversight, increased staffing, and infrastructure investments. But disasters can be catalysts for change, and I read about all these scares and mishaps with hope the aviation industry might finally pay attention to its underlying problems, which COVID helped reveal. Perhaps the ripples set off by a virus that killed millions of people worldwide would end up saving lives. Perhaps.

Ever since Cesare's crash, I'd awakened to the safety short-falls of the aviation industry. After solving the mystery of TACA 390, I tracked such shortcomings within aircraft manufacturers, airlines, regulatory agencies, and infrastructure. And I'd been inspired with a newfound purpose: I couldn't save my brother, but I could leverage him, his story, to try to save other lives. All these high-profile near misses, technology failures, and workforce issues occurred around the time the FAA Reauthorization Bill was up for approval by Congress. The five-year bill would renew funding for FAA operations, staffing, and infrastructure investment for improvements to runways, taxiways, and terminals. The bill also requested additional monies to address increased safety measures such as runway warning and alert systems, the kinds that might have prevented the rash of recent near misses. But the bill was stuck in a very divided Congress that didn't seem to agree on much. The FAA Reauthorization was set to expire September 2023 and approval before then seemed highly unlikely. The bill would require an extension through the end of March 2024 while our legislators hashed out their differences. This was my chance to use my voice to advocate for its passage.

As a board member of the American Society of Civil Engineers, I had added my voice to raise awareness about the condition and safety of various infrastructure types, including the nation's aviation facilities. After giving the US aviation's infrastructure their D+ in the ASCE's Report Card for America's Infrastructure, ASCE estimated almost $250 billion over ten years would be needed to address the shortcomings in our nation's airports.

These estimates were mind-blowing considering President Biden's Infrastructure Investment and Jobs Act, intended to upgrade airport facilities, would provide only 10 percent of the needed investments. The other 90 percent would need to come from Congress via FAA Reauthorization bills renewed every five years.

On February 27, 2024, I boarded an American Airlines flight to Washington, DC, where I was scheduled to join an ASCE contingency of fellow advocates. I had upgraded to the exit row because ever since TACA 390, I wanted to be part of the team helping to lead evacuations in the event of a flight mishap. The laminated trifold card in the seat pocket informed me I was flying on an Airbus 320. Luck of the draw. I had quickly learned the impracticality of avoiding Boeing planes. With an aircraft manufacturing duopoly consisting of Boeing and Airbus, there weren't many flight options available by limiting myself to Airbus planes. But on this day, I was lucky.

As a member of the ASCE contingency, I was headed to the halls of Capitol Hill to advocate for speedy passage of the FAA Reauthorization bill, whose extension was due to expire at the end of March. As is the case with so many things in Washington, the FAA bill was stuck in the convoluted legislative process as legislators debated several elements of the bill. I had participated in advocacy trips before, but this time it was personal. I had a story to tell. For the first time, I would share my account of the TACA Flight 390 tragedy with legislators in an advocacy setting.

The pilot announced we'd soon be landing at Ronald Reagan Washington National Airport. I smiled to myself as I recalled the article on the airport's website that described the construction work underway to repair the pavement failure on its main runway, the busiest runway in the country. Repair work had forced a temporary runway closure resulting in flight delays for thousands of travelers, including several legislators negotiating the bill. Reagan National's runway pavement failure was the very same type experienced at Toncontin Airport that was followed by shoddy repairs. Passage of the bill would ensure adequate infrastructure investments intended to prevent these types of pavement failures. When meeting with legislators, this would be my most convincing talking point: Pass the bill, and the funding will ensure we maintain safe runways and avoid delays. What I'd discovered in the many years working in the public sector was that sometimes we can be most persuasive by leveraging happenstance. Experiencing delays due to crumbling runways can be a great motivation to legislators as they fly back and forth between DC and their home states. Sometimes, the infrastructure gods act in mysterious ways.

Two days later after landing in DC, I sat with Representative Grace Napolitano in her congressional office. The walls were lined with photos and awards reflecting a life in public service at the federal, state, and local levels. I was a little nervous and somewhat starstruck. Congresswoman Napolitano, with her silver hair and many years of experience, was one of a few Latina women in Congress and since 2010 had served on the House Committee on Transportation and Infrastructure, which oversees the US aviation system. As a Latina woman myself, I was always in search of women that served as catalysts for change, and Representative Napolitano was one of my sheroes.

Representative Napolitano smiled at me. "I'm retiring this year," she said. "It's time."

"Well, I thank you for your service," I said. "But before you go, I'd like to ask for your support of the FAA bill, which is making the rounds through the House now."

"Of course, aviation safety is so critical. Every day there's a story about some mishap."

"My brother died in a plane crash," I blurted out. "It was preventable. The runway was poorly designed and barely maintained. And just like what we're seeing today, there were many warning signs. But nobody did anything about any of them."

"Oh my. I am so sorry," she said, her eyes clouding over. "We will pass the bill. You will see. And it will be in a bipartisan decision."

I had just met her not twenty minutes earlier, but I felt like reaching out to hug her. Just before I left her office, she handed me a congressional challenge coin engraved with her name and the Great Seal of the United States and said, "You have turned a tragedy into a mission to save lives. Keep up the good fight. Thank *you* for your advocacy." Then she pressed the coin into my palm.

I beamed with pride. I'd used my voice and Cesare's memory to advocate for aviation safety at the highest levels of government. And I'd been heard.

As I navigated my way through the crowded underground walkways connecting the maze of federal buildings, I caught sight of Senator Jamie Raskin of Maryland. He was standing in the middle of the wide tunnel engrossed in something on his phone. I knew Senator Raskin had lost his son to suicide and that he himself was a cancer survivor, a full head of curly brown hair just recently grown back. I'd practiced a thirty-second pitch, so now I seized the moment. "It's a pleasure to meet you, sir," I said, extending my hand.

Senator Raskin looked up a little startled but shook my hand and flashed a cheek-to-cheek smile. "Good to meet you," he said.

"My name is Rossana D'Antonio, I'm a board member of the American Society of Civil Engineers, and we're here in Washington advocating for safe, resilient infrastructure. Specifically, I'm asking for your support for the passage of the FAA Reauthorization bill. I have a vested interest because my brother died on a faulty runway."

"Oh, I'm so sorry. Yes, I fully support the FAA bill, and we're working hard to get it through the finish line," he said.

"Thank you! I'll let you get back to your work," I said, pointing to his phone.

"Good luck on the rest of your congressional visits," he said, shaking my hand again.

I dashed across the street to the Hart Building for my 3 p.m. appointment with Senator Alex Padilla. He'd been called away to another meeting, so I was greeted by Sam Mahood, his policy advisor. We were seated in a bright conference room around a rectangular table. I'd met Sam before on prior legislative visits, but this year, I was here to tell my personal story. "As an engineer himself, I'm sure Senator Padilla understands the need to pass the FAA bill quickly. I know firsthand the importance of investing in safe and resilient infrastructure," I said. "I lost my brother due to a flawed airport runway. He was a pilot, and his name was Captain Cesare D'Antonio. He died in the line of duty on a cold, stormy morning while attempting to land on a runway that everyone knew wasn't safe. None of the responsible parties had done anything to prevent it from happening. It happened to my family, and it can happen again. This tragedy would have been avoided had there been strong leadership to do the right thing."

Sam shifted in his seat. "I am so sorry. I appreciate your willingness to share your personal story. It's a sobering reminder to us all that the need to invest in our infrastructure is really about investing in our safety and quality of life," he said softly. "Senator Padilla supports the bill. He'll fight to ensure it passes."

Everyone I met with had an immediate reaction to my story. My brother was no longer a data point. My being there showed that passing safety legislation isn't about measuring aviation safety as statistics in a report; I was there to tell them the name of one person who died in an airplane accident. I felt seen and heard. I was taking care of my little brother.

The aviation industry prides itself on the notion that air travel is the safest mode of transportation. The data confirms this. The traveling public understands this, and there hasn't been a fatal commercial air crash since 2009 when Colgan Air 3407 crashed into a home in Buffalo, New York, killing all forty-nine passengers and the homeowner. Fifteen years with no catastrophic incident is a very good track record, but the events of 2023 alone highlight our complacency, a hubris in our greatness. All of us. Airlines, regulators, and passengers alike. So many near misses met with apathy and inaction. It reflects a culture of accepting unnecessary risks.

The events of these last two years highlight a defining moment for the aviation industry. It's time to be proactive about prevention instead of sticking our chests out and boasting about no recent deaths. When interviewed regarding the rash of near misses, Sully said, "We can no longer define safety as the absence of accidents."

The complex aviation system is under strain, and we have a chance to defuse the ticking bomb before it blows. If we don't, there will be more near misses. And inevitably, something very bad will happen, and we'll ask ourselves why no one did anything to prevent something very bad from happening.

That night, I strolled the short distance from my hotel to the National Mall. It was a crisp evening with not a star in sight. I was scheduled to fly home in the morning, so I took my time reliving the events of the day. I arrived at the steps of the Capitol flanked by the grandness of historic buildings that are the cornerstone of our nation's democracy. A government by the people, people like me whose responsibility it is now to advocate for those that cannot. As I stood amongst the fully illuminated structures it was hard not to be swept up in the meaning of the day of advocacy I'd just been a part of. Perhaps today it had been my turn to be the butterfly, flapping my wings in hopes of change.

22

Peace

Other than our parents, our siblings provide us with the longest lasting relationship we will know. My relationship with Cesare began when I was four years old and my mother presented me with a baby brother, swaddled in a blue blanket. At that moment, my life changed as I eased into the role of protector. Sentinel. Guardian.

Through the years, Cesare and I shared the death of a father, soared through the heavens, and found our bearings as we embarked on adulthood. And I was there to bid him farewell at the very end.

BlueMountain.com continues to send me email reminders for his birthday—the first one arrived just months after he was gone. *Cesare's birthday is December 15! Wish him a happy birthday!* I stared at the email for a long time, feeling nothing, aware I'd never again be able to wish my brother happy birthday. My brother had been robbed of years yet to be lived. The promise of what could have been. A future that would never be written. Cesare would remain forever young.

My mission for truth delivered the answers I needed, but it also reignited my determination to work toward ensuring safe and resilient infrastructure with a focus on aviation safety. The needless death of my brother fired me up even more than before to be a crusader for the public's well-being. I knew in time there would be another natural or human-made disaster, and I vowed to stand ready to respond using all the skills and experience I have. We can literally save lives through advocacy and public policy, so this cause is now my mission. I owe this bolstered sense of purpose to Cesare.

But even though the end of my investigation gave me inspiration, I realized I needed time to heal. I found healing in nature, specifically in Malibu, California.

It was Valentine's Day 2010, and Freddie and I were in our BMW, the sunroof open, my hair dancing in the wind, heading west through Topanga, past its tiny shops and bohemian, hippie houses, rustic homes with eclectic and colorful artwork in the front yards. The ride offered a sharp contrast to my conservative lifestyle that included business suits, a leather briefcase, and a four-door German sedan. It felt liberating.

Cradled by the steep canyon walls blanketed with thick brush and dense oak trees, I allowed myself to absorb the beauty of the area for the first time. Before Cesare's death, these canyons had simply offered a means to a destination for me; I'd driven them countless times for work while performing engineering inspections and determining the stability of mountain slopes. I'd always been in a hurry and had never stopped to appreciate the explosions of color of bougainvillea vines or the rainbows of wildflowers. I could see now that in my haste, I'd been missing all this wonder. Maybe Frank Lloyd Wright was on to something when he declared, "I believe in God, only I spell it *nature*."

The canyons of Malibu are filled with residential extremes. There are enormous mansions and little shack-type homes notched into the mountains and inconspicuously hidden behind the thick overgrowth. This area had a well-known history of canyon fires that devastated entire communities, and just when a fire season was over, a community might be subjected to heavy flooding and road washouts. There was also a chance rockfalls would cut off access to the residents from the rest of the community and hinder emergency services such as ambulance or fire response. Just as I'd once pondered the allure of flying despite the risks, I often wondered what inspired people to live in the middle of nowhere, always having to brace for the next natural disaster. But as I began driving to Malibu for pleasure, I began to understand what attracted people to the area. Malibu's stillness and wild beauty offered a wonderland on the edge of the big city.

A regal red-tailed hawk soared elegantly over the fog-filled hollows of the canyon bottom. Its outstretched wings spanned several feet, and at times the great bird hovered overhead as if escorting us, as if taking a moment to admire our surroundings as we did. Freddie and I took long, slow breaths, appreciating that we'd entered another world, a place where serenity governs, interrupted only by the soft rustling of the leaves in the crisp afternoon breeze.

At the bottom of the canyon, we turned right onto Pacific Coast Highway and continued to drive parallel to the blue waters of Malibu. With no destination in mind, I rolled down the windows and let the salty sea air fill my lungs. We found a nice clearing and parked along the side of the road. There we were, less than an hour from our home, yet it seemed we'd entered a magical place where the back-and-forth rocking of the waves offered a lullaby.

Freddie and I walked hand in hand toward the water and sat on the warm sand in a remote part of the beach embraced by the cool afternoon breeze and blanketed by the sun's rays. The glistening waters sent gentle rolling waves to tickle my toes, retreating quietly and leaving behind a carpet of effervescent foam.

It was on that beach I realized there was no clear distinction between where the ocean ends and where the heavens begin, between where I sat and where I imagined Cesare now soared. As if on cue, a lone wispy cumulus drifted into sight. Nearby, a couple of seagulls soared overhead. One was graceful and glided over the intense blue backdrop as the other one played in the ocean waters as if performing a dance just for me.

Fog rolled in as the sun began to set, and a cool afternoon breeze gently shooed us off. Sitting there, I realized the chest tightness that had been part of my every day since Cesare's death was gone, washed away in the grains of sand around me. I decided then I'd return soon to this magical place, and I could almost hear the gentle rolling waves whisper back at me, "We'll be waiting."

Both my brother's life and his death reminded me that all we have is today. Whether Cesare was riding the waves on the tropical beaches of El Salvador, bungee jumping off a bridge in New Zealand, or greeting the dawn of a new day in Greece, he'd always led his life with exuberance and joy. Inspired by my brother's gusto and his love of nature, two years after his death, Freddie and I moved to our dream home in the paradise of Malibu. It was here that I discovered peace and learned to live again.

Cesare was an inspiration in life. Cesare is an inspiration after death. He will forever be a gift.

23

Not in Vain

He who saves a life, saves the world entire.
—Talmud

This was the anonymous, handwritten note my family received a few days after Cesare died. To this day, I don't know which compassionate person provided such balm for the soul, but my family remains grateful for the message, which we assumed was about actions on the day of the crash. There were 132 people on TACA Flight 390 who walked away from the broken fuselage. Several others in surrounding businesses and homes were spared when the plane was deflected away from them. But time would prove how prescient these Talmud words would be.

A lot has happened as a result of my brother's plane crash on May 30, 2008. The most obvious change is that TACA Airlines, the flag carrier for El Salvador, is no more. In October 2009, TACA Airlines announced its intent to merge with the Colombian-based Avianca Airlines. That was about the time I last visited TACA headquarters and noticed signs of their rebranding campaign. Clearly, the strategy was just the beginning of much bigger changes to come. The merger was completed in May 2013, with TACA retiring its name and adopting the Avianca branding and logo. With the airline's demise, the TACA 390 tragedy seemed to be fading into the history books.

In a 2009 coup orchestrated by the Honduran army, President Manuel Zelaya was ousted, his successor supported by private business owners. Zelaya's decision to temporarily close Toncontin International Airport following the crash didn't bode well with the business community and may have contributed to the derailment of his political career. Despite that he was outspoken about blaming the pilot of TACA 390, he also recognized the need to permanently close deadly Toncontin, and he committed to relocating operations to a safer airport, effective 2009. But his plan was halted when he was removed from office. Negotiations for an airport relocation plan quietly continued for over a decade.

During this time, critical safety improvements were finally made at Toncontin. Massive grading was performed near the threshold of runway 02 to remove a large part of the dangerous hillside. In 2010, the asphalt runway was extended by 984 feet, and runway lighting systems were improved. These were the very first impactful runway improvements made to Toncontin since 1948. Despite these long overdue improvements, there were two additional small plane crashes, in 2011 and 2018. Between the two crashes, fourteen people were killed and six were injured. Both aircraft were attempting to land.

When people in power refuse to take note of glaring evidence about systemic failures in our built environments, history is bound to repeat itself. Toncontin Airport was one of the most dangerous airports in the world, so much so that in 2010, the History Channel highlighted it as one of the most treacherous airports in its "Most Extreme Airports" documentary series. Toncontin was rated second most dangerous, surpassed only by Nepal's Lukla Airport, a small airport accessible only to small fixed-wing aircraft. Landing at Lukla requires a visual approach on its short, 1,729-foot runway. It is the airport most climbers fly into as they embark on their climb to Mount Everest.

In 2016, a public-private partnership was established between an investor and the governments of Spain and Honduras to build a $163 million airport that would finally replace Toncontin. Finally, in November 2021, more than thirteen years after the TACA 390 disaster, Honduras inaugurated the Comayagua International Airport, also known as the Palmerola International Airport, located on the military base of Palmerola, which had served as the temporary airport following the TACA 390 crash. All international air carriers now use the new facility. Toncontin Airport is now restricted to local flights involving aircraft transporting no more than thirty-three passengers.

As is often the case with change, the airport switch was met with controversy. Citizens of Tegucigalpa argued that their commute to the airport would now be much farther, much less convenient. I was stunned by these protests, which served to make clear how short-sighted and uninformed consumers can be. We'll fight for convenience over our own safety. Surely, part of the reason for the public outcry was a short collective memory, which is one reason why after a tragedy like a plane crash, we need much faster investigations followed by swift changes.

When I heard that never again would commercial jumbo jets be allowed to land at Toncontin Airport, all I could think about was Cesare. Though he'd never know it, his death had led to changes that would save countless lives. *He who saves a life, saves the world entire.*

When Cesare, Mars, and I were kids, my dad used to tell us, "Live as if you can change the world." Cesare actually did. At least at Toncontin, history would not repeat itself. Cesare never set out to be a hero. But that's what he is.

24

Free

Protect your litter brother. Always protect your little brother. I had accepted that mighty mission when I was just four years old and had lived my life guided by this North Star. This was a heavy load for a kid to carry and unrealistic for anyone. As an adult, I wondered why my father didn't recognize that there's nothing any one of us can do to protect another. History proved I couldn't protect Cesare. But maybe that wasn't my father's point. Maybe that was never the point.

As an adult, I now understand people's responsibility to other people, to each other. Maybe it wasn't just about protecting my brother but also my brethren. Maybe I hadn't failed. Maybe my work was just getting started. Maybe.

It's clear to me that my father's death defined the course of our lives. Mars sought safety and security in marriage at a young age. I began reaching for a life of duty and caution. But our father's death affected Cesare quite differently. Recognizing at an early age that a life could be cut very short, he chose to make each day a grand adventure.

As Cesare's older sibling, I'd always assumed I should be the one to lead the way. But as it turned out, my little brother, both with his life and his death, has served as a guide for me. He's taught me not to sacrifice joy for fear of "what if." He's taught me that accepting risks and uncertainties is part of a fully lived life.

As Cesare did, I choose to trust the systems in place all around us. I have to trust in order to drive and fly in airplanes. In order to live in a naturally beautiful place, I have to trust that I'm as well prepared as I can be to face natural disasters. I must trust that my voice can have a positive impact in the highest levels of government.

Shortly after Cesare's death, I wished I could be transported into the future—even just one year ahead, to 2009. One year, I thought, would be enough time to spare myself the awful pain I was in every day since hearing of my brother's death. But I soon realized that there are no shortcuts, there's no turning back. What irony, that my peace would be found only by going through the pain.

I'd read Mary Oliver's work as I searched for comfort, and her words resonated as I stumbled along this path. I found one passage that was particularly meaningful:

> there was a new voice
> which you slowly
> recognized as your own,
> that kept you company
> as you strode deeper and deeper
> into the world,
> determined to do
> the only thing you could do—
> determined to save
> the only life you could save.

During one of my family's visits to Cesare's grave, I took my brother a gift. It was a decorative wind spinner in the shape of a Cessna plane, much like the plane he flew when I accompanied him to San Jose all those years ago. Freddie and I assembled it and drove the stake into the grassy area around the brass plaque. The little plane attached to a rod inserted in the stake, and when the wind blew, the plane's propeller spun, and the little plane turned on its axis.

The sun warmed us, and we were embraced by a crisp afternoon breeze. While Mom, Mars, Freddie, and I stood gathered at the foot of the grave, the little plane's blue-and-white propeller spun in the soft breeze. I couldn't help but smile. I was positive Cesare approved.

It was time to go, and we turned and began to slowly walk away. I looked back over my shoulder and saw the little plane had turned slightly on its axis and now faced us directly, its propeller spinning. I hung back, allowing the rest of my family to continue ahead down the walkway. As I watched, the spinning stopped. As I continued walking, the plane almost appeared to be responding to me as it continued to turn slowly on its axis. I stopped and it stopped. I waited to see if the breeze would set it in motion again, but it didn't. I took another two steps forward, and the plane turned slightly in my direction with the spinner going. I stopped, and the propeller stopped spinning.

I wondered, *Could it be?*

No, it couldn't.

Could it?

How could I even consider such improbable ideas? *I'm* the family skeptic. I looked around trying to gauge whether the breeze was actually blowing. It was blowing alright, but Cesare's little spinner remained still until I started walking again, and the little spinner continued to follow me—all the way to the cemetery's entrance.

My eyes watered, and I didn't want to leave. I wanted to stay and play with Cesare. I wanted to run up and down the walkway to see if the spinner would follow me. I wanted to run behind it and have the Cessna spinner twirl on its axis as it looked for me. I wanted to chase it screaming, "Olly olly oxen free!" at the top of my lungs. I wanted to hide behind the tall, twisted trees and lunge out yelling, "Boo!" I wanted to stretch my arms wide and twirl until I was so dizzy I collapsed on the lush green bed of grass.

Instead, I stood and watched the spinner turn wildly. I knew it was absurd, but somehow, I knew Cesare was happy—wherever he was. I could sense he was watching and playing. He was telling me it was okay. He wanted us to move on. Cesare wanted us to be free.

I stood at the end of the walkway and watched the propeller spin. I felt sure if it hadn't been held down by the rod it would have taken flight. It would have tilted upward and glided high and out of sight, soaring into the blue heavens above. Free. Just like Cesare. Forever free.

Acknowledgments

Writing this book has been the hardest thing I have ever done. Along this fifteen-year journey, there have been countless people who have been instrumental in helping me get this labor of love out into the world and into your hands. Perhaps the most pivotal influence has been my brother, Cesare, who served as my muse. On days when I thought this project would never take flight, Cesare would peer at me from his picture frame nestled amongst my most prized possessions on the bookshelf in our home office. The snapshot, taken in an Airbus 320 cockpit, is of a uniformed Cesare seated at the captain's chair sporting his Prada sunglasses, his crooked smile displaying a chipped front tooth. He is forever in his happy place, at 35,000 feet over Mother Earth, and he would nudge me to keep at it. I could almost hear him reminding me this story needed to be told. I thank him every day for being my inspiration, my teacher, my pesky little brother. I miss you every single day.

Thank you to the UCLA Writers Extension Program that introduced me to a vibrant writing community that helped shepherd this project forward. It was this tribe of magnificent storytellers that transported me into inspiring worlds of wonder.

Infinite gratitude to Barbara Abercrombie for years of inspiration as the ultimate hostess of writer's retreats in the peaceful and idyllic setting of Lake Arrowhead where I first discovered the healing power of nature. It was Barbara who first proclaimed me a writer as she intently listened to me read an excerpt of my initial draft, her hand over her heart, her eyes a well of tears. I wish you were here to see this.

A million thanks to Jennie Nash, book coach extraordinaire. Jennie was the first to hear the raw story. In those first few months following the crash, she helped me tap into the pain and formulate my thoughts, and through the years, she challenged me to dig deep and discover the bigger story. Jennie was absolutely right, fifteen years needed to pass in order for me to be able to write the story that needed to be told. Thank you for believing in me.

I am indebted to Jodi Fodor, my rockstar editor, for her partnership in what seemed like an endless cycle of edits and revisions. I could always count on her kind honesty and gentle nudging that forced me to search deep within me for uncomfortable answers. Jodi helped shape and mold this memoir into the best version possible.

Brooke Warner, my superstar publisher, you are a dream maker and have helped revolutionize the industry. Thank you for making this dream come true!

To the She Writes Press team, thanks for being a part of the movement and helping to shepherd this story forward.

I'm grateful to Jessie Glenn and the entire MindBuck Media publicity team. Thank you for loving this story and helping to spread the word.

I would be remiss if I didn't recognize the countless friends and family members on whose shoulders I leaned on many occasions. There are too many to mention by name herein, but I celebrate you and our friendship every single day.

I can't forget Luna, my fur baby, for accepting me as I am, diffusing my heightened stress levels, and lifting me up with her sloppy kisses and wiggly embraces!

To my sister Mars, *mille grazie per tutto*! We have been through a lot, and we are still standing, stronger and more resilient. Of the original D'Antonio *bambini*, it is now just me and Mars, divisible by no one. She is now my longest lasting worldly relationship, and I can't imagine this journey without her. I love you immensely.

Finally, much love to my rock, Freddie. This book would not have been possible if not for his endless patience, gentle guidance, and strong hugs. Thank you for being my staunchest supporter, my technical partner, and dare I say, my emotional balance. It was because of you that I learned to open my heart and allow myself to feel. Love you to the moon and back.

About the Author

Rossana D'Antonio is a licensed engineer with expertise in infrastructure design and emergency management and a strong advocate for infrastructure investments, including the Federal Aviation Authority Reauthorization. Shaped by her Italian and Salvadoran parents, D'Antonio is deeply committed to family which serves as the cornerstone for her commitment in creating an environment in which everyone feels safe and protected. She finds peace in Malibu, California, where she resides with her husband, Freddie, and puppy, Luna.

Looking for your next great read?

We can help!

Visit www.shewritespress.com/next-read
or scan the QR code below for a list
of our recommended titles.

She Writes Press is an award-winning
independent publishing company founded to
serve women writers everywhere.